AF876066

The Geologic Column

Perspectives within Diluvial Geology

John K. Reed and Michael J. Oard, Editors

Creation Research Society Books
6801 N. Highway 89
Chino Valley, AZ 86323

ISBN: 978-0-940384-34-7

Printed in the United States of America

Cover photo courtesy of C.R. Froede, Jr.

Contents

Preface

Geology forms a dividing line between Christian critiques of the worldview of naturalism. Most who affirm a theistic framework reject the atheistic conclusions of either gradualist or punctuationist evolution—rightly noting their incompatibility with a Christian worldview (or any theistic worldview, for that matter). Although some Christians continue to deny the pervasiveness of naturalism in other facets of the sciences and may even affirm theistic evolution, that perspective appears in decline due to serious theological and scientific challenges from the modern creationist and Intelligent Design schools.

But surging skepticism about Darwinian biology is not accompanied by a strong reaction against uniformitarian geology, which constructed the framework within which Darwin's species originated. Many Christians, satisfied that the inherent atheism of evolution has been blunted, have yielded the field, believing that the debate in geology is nothing more than a trifling detail about the age of the cosmos and the Earth.

However, recent decades have seen the rebirth of interest in diluvial geology. Although research is limited by resources, the past years have witnessed the accumulation of a serious body of literature addressing the biblical and geological implications of a history that includes the global destructive cataclysm of Genesis. Numerous uniformitarian shibboleths have been revealed as contrary to sound science and sound thought. Though secular neocatastrophists strain to deny any influence of the new diluvialism, it is more than coincidental that their movement has flourished in the wake of a renewed emphasis on the importance of the Genesis Flood.

As with any new paradigm, modern diluvial geology is the home of a remarkable diversity of original thought, with vigorous discussion on topics ranging from plate tectonics to the nature of the Ice Age. One of the most interesting debates is concerned with the relationship between the Genesis Flood and the uniformitarian geological timescale, or geologic column. What exactly is it? Are parts of it separable from the old-earth geochronology? What would a diluvial stratigraphy entail?

This symposium attempts to capture the current range of diluvial opinion about the geologic column in an effort to better clarify what has become an admittedly murky debate. Recent exchanges in the literature show a tendency for various participants to talk past one another, and so this book has brought those disparate views together to better define the real differences within that community and identify concrete issues that will provide the basis for continued research by adding clarity to the problems and their potential solutions. It is not an apologetic to uniformitarian geology; rather it assumes the reader will either be a diluvialist or will at least possess both an interest in the topic and an open mind towards its approach to natural history.

Though it may be hard to see at this stage of history, we are convinced that diluvial geology represents a major paradigm shift that holds the potential to stimulate a revolution within the earth sciences—a change in perspective that will cast off the Lyellian straitjacket that seeks to prevent geologists from investigating the greatest geological event in our planet's history. Such a revolution would—we are convinced—spur scientific rigor, restore a solid philosophical foundation beneath both science and history, and promote an increased confidence in biblical history and Christianity—the same religious culture that gave birth to modern science. We invite the reader to look and see.

John K. Reed
Michael Oard

one

Introduction

John K. Reed
Michael J. Oard

Road cuts on highways. Scenic overviews in the mountains. Hiking trails in canyon country. We cannot travel from place to place without noticing rocks. In one place we see flat-lying sediments in orderly alternating layers. In another, great blocks of granite rear up before our vision as rugged mountains. Clues to the titanic forces acting on the crust are unveiled in faults and folds. Limited to Earth's surface, most people see only tantalizing hints of the complexity of what lies beneath their feet. In recent centuries, geologists have come to appreciate that ordered complexity; curiosity and economics have combined to spur investigation into the mysteries of Earth's crust. Technology has multiplied our ability to see: in 1869, John Wesley Powell had to brave the dangers of the Colorado River to see the strata of the Colorado Plateau revealed in the sides of the Grand Canyon. Today, a geologist can sit in the comfort of an air conditioned control room, collect the detailed readings of instruments ascending a 30,000-ft well drilled in deep-ocean waters and "see" every inch of the penetrated strata while determining lithology and bedding; measuring density, porosity, and radioactive emissions; discovering whether oil or gas might reside in the pore spaces and then calculating how rapidly it might flow to the surface.

Through these efforts, we are continually gaining a better empirical understanding of Earth's crust. Geologists have described igneous basement rocks and the sedimentary fill of overlying basins. Folds, faults, and tectonic fabrics have been discovered and mapped, from roadside outcrops to continents. Both the three-dimensional geometry and the internal composition of rock bodies have provided means of dividing, classifying, and correlating these units. However, empirical data are merely building blocks. Their arrangements are determined largely by theoretical models; some are amenable to modification by the data they seek to explain, while others rest on untestable assumptions about nature and time. As has been true throughout human history, it is the conceptual clash of those indemonstrable axioms that tends to create the largest intellectual explosions.

In 1815, William Smith published what most consider the world's first modern geologic map, one that shows the rock record of England and Wales (Winchester, 2001). One of his most significant innovations was linking the lateral distribution of rock bodies perpendicular to strike into a superposed vertical sequence seen in his extensive explorations of pits and mine shafts. Since that time, most geologists have conceptualized the rock record in terms of a vertical succession of layers, examples of which are seen in the Grand Canyon. Each layer is thought to represent a discrete time period that can be correlated all around the Earth. *Therefore, the essence of the geologic column is an interpretive framework for the rock record built on successive, globally-correlated, synchronous time periods.*

Another complementary paradigm shift was underway in early 19th century Britain that dramatically reinforced this new way of looking at the rock record. The rising influence of the Enlightenment elite was beginning to call into question the biblical chronology and downplay the geologic effects of the Flood. James Hutton brought this fight into the geological arena at the end of the 18th century with his uniformitarian theory that added long periods of time to the equation—a view cemented into modern geology within a few decades by Charles Lyell. Geologists were faced with an enticing explanation for the rock record—it represented a vertical succession of time units, each displaying long periods of gradual sedimentation and other geologic activity. Since that time, the position that attributes the majority of the rocks to the action of the biblical Flood has been a minority position—at times, vanishingly small.

This new explanation meshed well with the exuberant confidence that science could unveil a universe long shrouded by religious dogma. The history recorded in Genesis traced a story of just a few thousand years. If the world's story actually stretched far beyond the oldest accounts into the mists of time, the only possible record of those lost ages would exist in the rocks. Thus, the vertical succession of rock layers presented itself as a great calendar, and the sedimentary layers became a chronological account of the newly-discovered "deep time." Furthermore, the fossil contents held forth the possibility of widespread correlation across disturbed zones and lithological changes, implying a global column. All that was required was to set the boundaries of these eras and then interpret the hidden stories in their constituent rocks. Since time had become the primary determinant in the new science of geology, and since geologic processes in the past were assumed to be no different from those observed in the present, it seemed reasonable that the different rock units would not only represent distinct time units, but that they would do so all over Earth. The millions of years available in each age would certainly allow local variations in sedimentation, tectonism, etc., to be averaged out into one representative section. This translation of rock units into globally correlative time periods was the key to uniformitarian stratigraphy—still the dominant school in historical geology today.

But in the world of ideas, there was to be no unending succession of such "uniform" thought. Just as the geologic calendar was being "calibrated" to purported absolute dates by radioisotopes, a few pioneers during the middle of the 20th century began to demonstrate unambiguous evidence of catastrophic events in the rocks. Over a period of decades, a school of "neocatastrophism" emerged which recognized that the rock record was punctuated by catastrophic events, but set them still within the context of a slightly revised history still dominated by uniformitarianism, evolutionism, and deep time. Coincident with that development was the late 20th century resurrection of the idea that the biblical Flood was a significant agent in geologic history. This new "scientific creationism" also mounted a spirited attack on the old ages that geologists had come to take very much for granted, and on the isotopic dating methods which supposedly took them from the realm of well-supported theory into that of inarguable fact.

In the biological realm, the concomitant assault on Darwinian evolution began to reverberate in geology, because stratigraphy had come to rely on the predictable ongoing, irreversible evolutionary development of life through time as the key that would unlock the code found in the fossil contents of the strata. Index fossils could supposedly be sequenced by their evolutionary progress, ensuring worldwide correlation to an irreversible, unending, global clock. Despite furious resistance, it has become clear that biological evolution has serious scientific weaknesses and its proponents have had to turn to the power of the courts to maintain their educational monopoly because the arena of ideas has become too inhospitable. However, the echoes of those weaknesses have not yet reverberated as loudly as they probably will within historical geology and stratigraphy.

Ironically, the resurgence of modern creationism is almost universally traced to the publication of *The Genesis Flood* in 1961—a book whose geological emphasis led to a biological revolution! Since then, the retreat of evolutionism has been steady, but the subject of the book—the Noachian Flood and its relationship to the rock record of geology—has taken hold much more slowly. There are a variety of reasons for the differential success of modern creationism with Darwinian biology as opposed to Lyellian geology. While many Christians easily see the threat that biological evolution presents to their worldview, questions about Earth's age and the meaning of the rocks seem to pose much less peril. Furthermore, while Christian scholars can question evolutionism and still maintain a semblance of academic respectability, the rejection of the geologic ages quickly places them outside accepted circles. and quite frankly, many theologians and church leaders simply do not understand geology well enough to comprehend the contradictions inherent in asserting biblical fidelity on one hand and the geologic ages on the other. Their theological background and focus on Genesis 1 and 2 have diverted them from the implications of Genesis 6–8.

However, there is another reason for the slower development of the new Flood geology: diluvial geologists have been unable to reach consensus on several foundational issues of stratigraphy and historical geology and thus make a unified and concerted argument for a new paradigm of geologic history. One of the primary areas of disagreement has been how to incorporate the crustal record into the history outlined in the accounts of Creation and the Flood found in Genesis. A survey of creationist literature of recent decades shows a gamut of opinion and a corresponding number of incipient models that have led to a tangle of debate—especially with regard to the geologic column.

The modern geologic column (or geologic timescale) is the marriage of relative and absolute dating methods to produce a global correlative standard for Earth's crust based upon the assignment of its component rocks to standardized time periods bounded by absolute age dates (Figure 1). It is a model that exercises significant conceptual power—being virtually unquestioned—but that possesses equally significant weaknesses in its assumptions and internal coherence. Furthermore, it is a concept of earth history that cannot be reconciled in its present form to biblical history for a myriad of reasons. Most of these will be explained and explored in the following chapters.

One of the biggest conflicts between the geologic column and biblical history is the imposition of an absolute timescale in excess of 4.55 billion years. Whether or not the relationship between the chronostratigraphic (relative) sequences and the geochronologic (absolute) dates is separable or not remains a question that is at the heart of the present diluvial debate. Uniformitarian scientists certainly suggest the possibility:

> The chronostratigraphic scale is an agreed convention, whereas its calibration to linear time is a matter for discovery or estimation (Gradstein, 2004, p. 3).

But does disagreement with the methods that purport to discover or estimate linear time also invalidate the chronostratigraphic scale, especially if it is no more than "an agreed convention"? That certainly seems to be the case for the Precambrian:

> ...Precambrian stratigraphy is formally classified chronometrically...the base of each Precambrian eon, era, and period is assigned a numerical age (Gradstein, 2004, p. 3).

In other words, the chronostratigraphic subdivisions of the Precambrian (i.e. Archean, Proterozoic) are arbitrarily defined by the geochronology—assignment of absolute dates. The Phanerozoic is more complicated. First, it must be recognized that different parts of the Phanerozoic are defined and calibrated by different methods:

> Aspects of the GTS89 [ICS Geologic Timescale 1989] compilation began a trend in which different portions of the geologic timescale were calibrated by different methods. The Paleozoic and early Mesozoic portions continued to be dominated by refinement of integrating biostratigraphy with radiometric tie points, whereas the late Mesozoic and Cenozoic also utilized oceanic magnetic anomaly patterns and astronomical tuning (Gradstein, 2004, p. 10, brackets added).

Of the various methods that have been historically used to construct Phanerozoic timescales, Gradstein (2004) emphasizes (1) biostratigraphy, (2) stratigraphic reasoning, (3) orbital oscillations, and (4) magnetic reversals. As noted above, the Paleozoic is defined by biostratigraphic control tied to radiometric "markers." Of course diluvialists reject both the possibility of an evolutionary fossil succession through time as well as the extended ages purportedly demonstrated by radiometric methods, though they remain open to a non-evolutionary succession representing an as-yet-defined order inherent to the rock record.

Moving up the timescale, diluvialists face similar problems; the absolute calibration of the timescale has become closely integrated with the chronostratigraphic definitions. Stratigraphers have begun linking cyclic properties in the rocks to orbital oscillations that demand millions of years:

> Cycle stratigraphy has calibrated the timescales for most of the Neogene Period (i.e. for the past 23 million years), and for portions of the Paleogene Period (from 65 to 23 Ma) and Mesozoic Era (from 251 to 65 Ma) [Gradstein, 2004, p. 4].

Furthermore, a similar relationship has been built between the rocks and their magnetic properties:

> ...oceanic magnetic anomalies are calibrated with spreading models to produce a powerful correlation tool for sediments deposited during the past 160 million years. These calibrated C-sequence and M-sequence polarity timescales enable assignment of ages to stage boundaries and to biostratigraphic and other stratigraphic events through much of that interval (Gradstein, 2004, pp. 4–5).

Clearly, the assumptions of uniformitarianism, the validity of plate tectonic models, the correlation of dissimilar oceanic crust and continental sediments by radiometric methods, and the contemporary causal links between astronomical phenomena and the rock record are unacceptable to diluvial geologists.

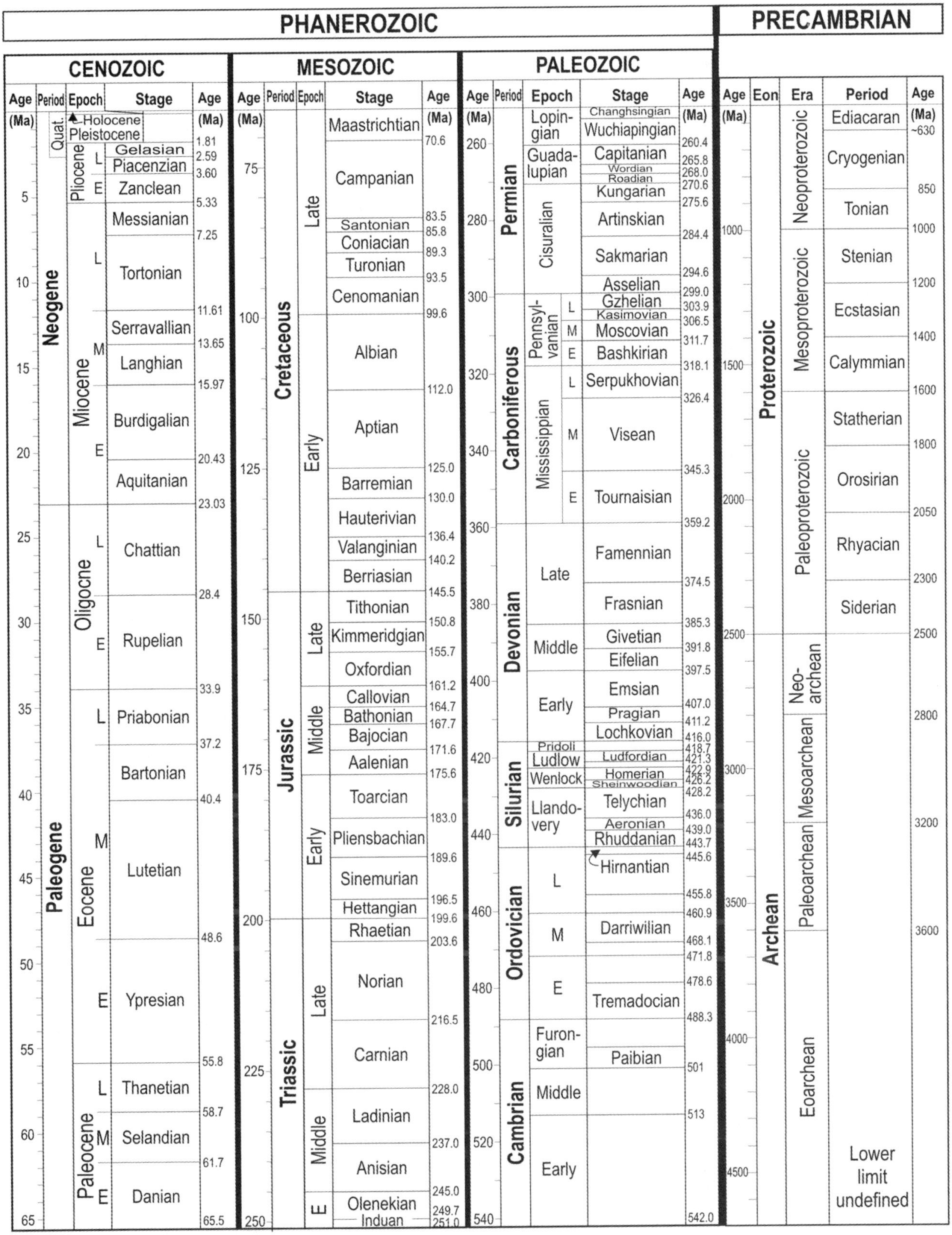

Figure 1. The geologic column. From Gradstein et al. (2004). Note that North American versions may differ slightly in terminology and convention.

So what remains? Can the chronostratigraphic skeleton be divorced from the geochronology and reinterpreted in the context of the Genesis Flood? Or is there a need to reject the entire paradigm that rocks in Earth's crust can be globally correlated by means of discrete time periods, regardless of their duration? Or is the answer somewhere between those two positions?

Given this broad disagreement over the rock record among those who accept a global flood as a historical reality, this book has been developed to provide clarity, encourage discussion, and promote progress in diluvial thought. In the confusion of debate, it is often profitable to step back, assess current arguments, and use that process to establish a new baseline for further discussion.

Such is the hope for this book. Chapters 2 and 3 will provide historical and technical background about modern stratigraphy—the discipline that is the gatekeeper of the rock record. Chapter 3 emphasizes agreement between diluvialists over the potential usefulness of the empirical rock record and many conceptual tools that can be used to investigate its features. The succeeding five chapters will present representative diluvial perspectives on the geologic column and supporting arguments. At the end of each of these chapters, a short section of questions and answers among the contributors has been added to stimulate much needed discussion. Finally, the book will conclude with a summary of the positions, noting areas of agreement and specific points of disagreement. While this book does not purport to capture the totality of modern diluvial thought, it does address major positions on the geologic column and cited references in its chapters will direct the interested reader to contributions that represent the full range of current opinion.

We hope that this symposium will better define essential positions relative to the geologic column and fairly present their strengths and weaknesses for all interested creationists. All of the authors are committed to the biblical account of Earth's history, including the unavoidable inference of a much younger world than is accepted by uniformitarian geologists. Furthermore, all of the authors are firm in their position that a global, catastrophic flood significantly affected the formation of what we today see as the rock record. If future diluvial geologists can point back to this volume as an accurate snapshot of diluvial views in the early 21st century (however confused they might appear at this moment) and see in it a point of departure towards the development of an integrated and coherent strategy for diluvial geology, then we will be well content.

References

Gradstein, F.M. Introduction. In Gradstein, F.M, J.G. Ogg, and A.G. Smith (editors). 2004. *A geologic timescale 2004*. pp. 3–19. Cambridge University Press, Cambridge, UK.

Whitcomb, J.C. and H.M. Morris. 1961. *The Genesis flood*. Presbyterian and Reformed Publishing Company, Philadelphia, PA.

Winchester, S. 2001. The map that changed the world. HarperCollins Publishers, Inc., New York.

two

The Historical Development of the Old-Earth Geological Timescale[1]

Terry Mortenson, Ph.D.

Abstract

This chapter will summarize the historical development of geology and in particular the development of the modern uniformitarian view of the geological record and the millions-of-years timescale. The important proponents of old-earth thinking (including Werner, Smith, Hutton, Cuvier and Lyell) along with some of the opponents of this view as it was developing (the scriptural geologists) will be discussed, as well as the key ideas in the development of the "geological column." This historical context is useful in assessing both the current evolution-creation controversy and the debate among creationists about the validity and role of the "geological column" in their models.

Introduction

The fundamental features of geological study, namely, field work, collection and theory construction, were not developed until the 17th to 19th centuries. Previously, back to ancient Greek times, many scholars believed that fossils were the remains of former living things and many Christians (including Tertullian, Chrysostom and Augustine) attributed them to the Noachian Flood. But other scholars rejected these ideas and regarded fossils as either jokes of nature, the products of rocks endowed with life in some sense, the creative works of God, or perhaps even the deceptions of Satan. In the 17th

[1] Much of the information in this chapter has been taken from various portions of Terry Mortenson's, *The Great Turning Point* (Green Forest, AR: Master Books, 2004) and is used here with the kind permission of the publisher.

and early 18th centuries the debate among naturalists intensified. One of the prominent opponents of the organic origin of fossils was Martin Lister (1638–1712), an English physician and naturalist. John Ray (1627–1705), the greatest naturalist of his time, favored organic origin although he respected Lister's objections. The debate ended when Robert Hooke (1635–1703), a British naturalist, confirmed from his microscopic analysis of fossil wood that fossils were the mineralized remains of former living creatures. However, he did not believe they were the result of Noah's Flood.

Early Geologists and Principles

Prior to 1750 one of the most important geological thinkers was the Danish anatomist and geologist Nicholaus Steno (1638–1686). Born to Protestant parents, he became a Roman Catholic in 1667. Steno was the first person to give a historical reconstruction of the formation of the geological layers of an area of the Earth, in this case the area of Tuscany, Italy, where he lived for some time. He did so within a biblical framework and argued that the geological evidence confirms the truth of the literal history in Genesis 1–11.

In his *Prodromus*[2] (1669), Steno expressed his belief that the Earth was almost 6,000 years old[3] and that organic fossils and the sedimentary strata were laid down by Noah's Flood. Based on his field observations in Tuscany and his reading about the geology of other locations, he set forth several principles for interpreting the rock record. Those principles laid the foundation for modern geology and are still taught and used by geologists today, although most of them probably have no idea that Steno was a Bible-believing, young-earth creationist.

Steno's key geological principles were these. One, most of the rock layers were once water-laid sediments. Two, marine deposits can be distinguished from fresh-water deposits by the fossils they contain. Three, the order of superposition of the layers indicates the relative age of the strata (the older deposited below the younger). Four, the sediments were originally deposited in an essentially horizontal fashion. Five, strata that are no longer horizontal were disturbed after deposition.

Shortly after Steno, Thomas Burnet (1635–1715), a theologian, published his influential *Sacred Theory of the Earth* (1681) in which he argued from Scripture, rather than geology, for a global Flood. He made no mention of fossils and though he believed in a young Earth he took each day in Genesis 1 to be a year or longer. Following him, the physician and geologist John Woodward (1665–1722) invoked the Flood to explain stratification and fossilization, in *An Essay Toward a Natural History of the Earth* (1695). Woodward suggested that the specific gravity of the sediments and fossils determined the order in which they were buried in the rock record. In *A New Theory of the Earth* (1696) William Whiston (1667–1752), Newton's successor at Cambridge in mathematics, shared similar views, though not based on personal geological investigation. But he suggested that a comet triggered the onset of the Flood and he added six years to Ussher's date of creation by arguing that each day of Genesis 1 lasted one year. Some of his reasoning was used later by those who favored the day-age theory for Genesis 1.

In *Treatise on the Deluge* (1768) the geologist Alexander Catcott (1725–1779) used geological arguments to defend the Genesis account of a recent creation and global Flood which produced the geological record. On the continent, Johann Lehmann (1719–1767), a mining and mineralogy professor in Berlin, carefully

[2] This was intended to serve as an introduction to a longer work on geology, which never was written. See the English translation and introduction by J.G. Winter in White (1968).

[3] He accepted Archbishop James Ussher's 4004 BC date of creation.

studied German mountain strata and recognized three classes of rocks: the non-fossil-bearing and greatly inclined rocks in the high mountains (formed during creation week), fossiliferous rocks resting roughly horizontally on the flanks of the mountains (attributed to the Flood) and loosely consolidated rocks on top of these (produced after the Flood by accidents of nature from time to time).

Development of the Old-Earth Theory

The idea that the Earth was much older than biblical history slowly replaced the traditional view during the late 18th and early 19th centuries. Geologists such as the Frenchmen Jean Elienne Guettard (1715–1786) and Nicholas Desmarest (1725–1815) and the Italian Giovanne Arduino (1714–1795) denied the Flood and advocated a much older Earth. In his study of mountains Arduino agreed with Lehmann in many ways and gave the three classifications of rocks the names *Primitive*, *Secondary* and *Tertiary*. The English geologist John Whitehurst (1713–1788) contended in his *Inquiry into the Original State and Formation of the Earth* (1778) that the Earth was much older than man and that although the Noachian Flood was a global catastrophe it was not responsible for most of the geological record.[4]

In France, three prominent scientists advanced the old-earth idea. In his *Epochs of Nature* (1778), which built on his thinking expressed in his *Theory of the Earth* (1749), Comte de Buffon (1708–1788) advocated the hypothesis that Earth had originated from a collision of a comet and the sun. Extrapolating from experiments involving the cooling of various hot metals, he postulated that Earth had passed through seven epochs over about 75,000 years to reach its present state. He had an evolutionary view of the origin and diversification of life (but offered no mechanism for such change) and his geological ideas were similar to Lyell's later uniformitarian theory. The astronomer Pierre Laplace (1749–1827) was strongly motivated to eliminate the idea of design or purpose from scientific investigations. As a precursor to modern cosmic evolution, he proposed the nebular hypothesis in his *Exposition of the System of the Universe* (1796). According to Laplace, the solar system was produced slowly over long ages as a hot, spinning gas cloud cooled and condensed by the laws of physics and chemistry. Jean Lamarck (1744–1829) was a naturalist specializing in the study of fossil and living shell creatures. In *Zoological Philosophy* (1809) he attempted to explain the similarities and differences between living and fossil creatures by a process of gradual evolutionary transformation involving the inheritance of acquired characteristics. He believed in spontaneous generation, rejected the notion of extinctions, and became a fierce opponent of the catastrophist Georges Cuvier.[5]

The years 1780–1820 have been called the "heroic age" of geology. During this time geology truly became established as a separate field of scientific study. More extensive geological observations began to be made, new methods were developed for systematically arranging the rock formations, and the Geological Society of London (the first society fully devoted to geology) was born. But it was also during this period that geology became embroiled in the so-called Neptunist-Vulcanist[6] debate (Gillispie, 1951; Hallam, 1992). Neptunists believed that all the rock formations were deposited by water, whereas the Vulcanists contended that granites and basalts were produced by molten lava being pushed to the surface from deep in the Earth. The leading advocates of the two positions were, respectively,

[4] For further discussion of these 17th and 18th century writers on geology, see Rudwick (1985, pp. 1–93) and Young (1988, pp. 27–42).

[5] For further discussion of these three writers, see Brooke (1991, pp. 234–242) and Hahn (1986, pp. 256–276).

[6] Neptunism was named after the Roman god of the sea whereas Vulcanism takes its name from the Roman god of fire.

Abraham Werner (1749–1817) of Germany and James Hutton (1726–1797) of Scotland.

Uniformitarianism

Werner was one of the most influential geologists of his time. He wrote very little. But as an excellent mineralogist and an inspirational and dogmatic teacher for 40 years at the University of Freiberg, Werner attracted the great loyalty of his students, many of whom became prominent geologists in the 19th century (Laudan, 1987). His theory of earth history was briefly explained in his otherwise descriptive, 28-page *Short Classification and Description of the Various Rocks* (1786). As a result of his study of the strata in Saxony, which were clearly water-deposited, he proposed that most of the Earth's crust had been precipitated chemically or mechanically by a slowly-receding primeval global ocean over about one million years. Building on the ideas of Lehmann, Werner divided the rock layers into five periods. The lowest three, deposited by the global ocean and constituting most of the geological record, were the *primitive* rocks (devoid of fossils), *transition* rocks (strata with some fossils and roughly equivalent to the lower Paleozoic), and the *floetz* formations (the fossiliferous strata from the upper Paleozoic through the Cenozoic). The upper two nearly-contemporaneous formations were the *alluvial* (relatively unconsolidated deposits) and *volcanic* and were the result of local depositional events. Although Werner's theory of earth history was attractively simple, many objections were soon raised against it. But the idea of millions of years stuck with his students.

Hutton's geological views were published first in a 1788 article and then in a two-volume book in 1795, both entitled *Theory of the Earth*. His theory was significantly different from Werner's. He argued that the primary geological agent was fire, not water. The sedimentary rocks were the result of detrital matter from a previous state of the world being slowly eroded off the continents and carried by rivers to the ocean floor. There the sediments were gradually transformed into rock by the Earth's internal heat and later raised by convulsions to form the land masses that we have today, which in time would be eroded into the oceans and later raised to become new continents.

Hutton's view was an early version of uniformitarianism: everything in the rock record must and can be explained by present-day processes of erosion, sedimentation, volcanism, and earthquakes occurring at a regular rate throughout earth history.[7] His cyclical view of history expanded the age of the Earth almost limitlessly. In fact, he asserted that he saw "no vestige of a beginning or prospect of an end" in the geological record.

In his *Illustrations of the Huttonian Theory of the Earth* (1802) John Playfair (1748–1819), mathematician and Scottish clergyman, republished Hutton's ideas in a more comprehensible form. He defended Hutton against the charge of atheism levelled by a number of contemporaries by arguing that Hutton's ceaseless cycles of geological processes were like Newton's laws of regular planetary motion. Although Playfair made no attempt to harmonize Hutton's theory with Scripture, he did defend Hutton's notion of the Earth's great antiquity by asserting that the Bible only addresses the timescale of human history, which Hutton did not deny was relatively short, as the biblical chronology indicates. Like Hutton, Playfair also argued that the Flood was tranquil, not a violent catastrophe.

Catastrophism

Neither Hutton nor Werner paid much attention to the fossils. In contrast, William Smith (1769–1839), a drainage engineer and surveyor, became fascinated with the fossils and rock layers in the course of his

[7] This was not a completely new idea; Aristotle expressed similar views in his *On Meteorology*. See Rudwick (1985, pp. 37–38).

work. After many years of field observations, he published three works from 1815 to 1817, which contained the first geological map of England and Wales and an explanation of the order and relative chronology of the stratigraphic formations as defined by certain characteristic (or index) fossils rather than the mineralogical character of the rocks.[8] He became known as the "Father of English Stratigraphy" because he gave geology a descriptive methodology for assigning relative ages to the various formations. This became critical for the establishment of the theory of an old Earth. Though Smith believed that a global flood was responsible for producing the gravel, clays and sands scattered over Earth's surface, he never explicitly linked this with the Noah's Flood and he believed that all of the sedimentary strata were deposited many long ages before this flood by a long series of supernaturally-induced, large-scale catastrophic floods separated by long ages. After each flood, God created of new forms of life to replace what was destroyed.[9]

In the early 1800s Georges Cuvier (1768–1832), the famous French comparative anatomist and vertebrate palaeontologist, began to develop his theory of catastrophism[10] presented in his *Theory of the Earth* (1813). It went through several English editions over the next twenty years, with an appendix (revised in each later edition) written by Robert Jameson, the leading Scottish geologist. The son of a Lutheran soldier, Cuvier sought to show a general concordance between science and religion (Coleman, 1973). In his *Theory* he seems to have treated post-flood biblical history fairly literally, but did not interact with the text of the scriptural accounts of Creation and the Flood at all. He reacted sharply against Lamarck's evolutionary theory and his denial of extinctions. From his study of large quadruped fossils from the Paris basin, Cuvier concluded that many species had indeed gone extinct, but not all at once. Rather, like Smith, he theorized that in the past there had been many catastrophic floods. and with Smith he believed that each stratum was characterized by wholly unique fauna. The fauna had appeared for a time, were catastrophically destroyed, and then new life forms arose. Although Cuvier rejected Lamarckian evolution, it was not clear whether the new creatures were supernaturally created after each flood or if the stricken area was repopulated through migration of similar animals from outside the flood zone. He believed that Earth history was much longer than the traditional 6,000 years, but that the last flood had occurred only about 5,000 years ago. This obviously coincided with the date of Noah's Flood, but Cuvier never explicitly equated his last flood with it.[11] These violent catastrophes were vast inundations of the land by the sea. But they were not necessarily global, therefore whole species were not always eliminated. Man had first appeared sometime between the last two catastrophes.

Another important development at this time was the establishment of the Geological Society of London in 1807. It was the world's first scientific society devoted solely to geology. The thirteen founding members

[8] William Smith (1815; 1816; 1817).

[9] See Phillips (1844, pp. 25–26) and Smith (1835). The latter work (a large one-page explanatory diagram) was Smith's last and clearest statement on his view of Earth history and was obviously intended to be a response to Lyell's uniformitarianism. Though, when he referred to the "Deluge" he possibly meant the Noachian Flood, he made no reference to Scripture. However, he was quite emphatic about the supernatural nature of the many revolutions and creations after these catastrophes.

[10] The terms "catastrophism" and "uniformitarianism" were coined by the historian and philosopher of science William Whewell (1832, p. 126) in an anonymous review of Lyell's *Principles of Geology*.

[11] It was the editor and publisher of Cuvier's English editions, Robert Jameson, who made the clear connection between Cuvier's last catastrophe and Noah's Flood, no doubt to make it more compatible with British thinking at the time. The Oxford geologist William Buckland made this idea even more popular. See Rudwick (1985, pp. 133–135).

were wealthy, cultured gentlemen, who began with little geological knowledge but quickly learned. From its inception it was dominated by old-earth thinking (the relation of Genesis to geology was never discussed in its public communications), although it did not overtly favor either uniformitarianism or catastrophism as those views were developing.

William Buckland (1784–1856), professor of geology at Oxford, was the leading geologist in England in the 1820s, initially following the catastrophism of Cuvier and Smith. Like many scientists of his day, he was an Anglican clergyman. Two of his students, Charles Lyell and Roderick Murchison, went on to become very influential uniformitarian geologists in the 1830s and beyond. In his efforts to get science, and especially geology, incorporated into the Oxford curriculum, Buckland published *Vindiciae Geologicae* (1820). Here he argued that geology was consistent with Genesis, confirmed natural religion by providing evidence of creation and God's continued providence, and proved virtually beyond refutation the global, catastrophic Noachian Flood. However, the geological evidence for the Flood was, in Buckland's view, only in the upper formations and topographical features of the continents; the Secondary formations of sedimentary rocks were antediluvian by untold thousands of years or longer. To harmonize his theory with Genesis he considered the possibility of the day-age theory, but favored the gap theory. He never worked with the text of Genesis to show how old-earth theory could be harmonized with the Bible. Like Cuvier, he believed in multiple supernatural creations and that the creation of man was only a few thousand years ago.

After further research, he published his widely-read *Reliquiae Diluvianae* (1823) providing what he thought was a further defence of the Flood (albeit limited in its geological effects). However, the uniformitarian criticisms of Charles Lyell and others led Buckland to abandon this interpretation of the geological evidence in the early 1830s. He publicized this change of mind in his famous two-volume *Bridgewater Treatise* on geology in 1836, where in only two brief comments he described the Flood as tranquil and geologically insignificant (pp. I:16, 94–95). It is clear from Buckland's personal correspondence in the 1820s that geological evidence had a superior quality and reliability over textual evidence in reconstructing the Earth's history (Rupke, 1983), because written records were susceptible to deception or error, whereas the rocks were truthful and could not be altered by man. He did not assert biblical fallibility, but certainly implied it in his statement.

Adam Sedgwick (1785–1873) was Buckland's counterpart at Cambridge, receiving the chair of geology in 1818. He too was an ordained Anglican clergyman and insisted all his life that old-earth theories did not contradict the Bible, but he never once attempted to show how they could be harmonized. Like Buckland, he was a catastrophist in the 1820s, but publicly recanted this view in 1831, embracing Lyell's uniformitarianism. Through the influence of these two clerical/academic geologists and others (e.g., George Greenough, Rev. William Conybeare, Roderick Murchison and Henry De la Beche in England and many geologists in Europe), old-earth catastrophist (or diluvial) geology was widely accepted in the 1820s by most geologists, and many clergy and theologians in Britain and North America.

There were several reasons most geologists at this time believed the Earth was much older than 6,000 years and the Noachian Flood was not the cause of the Secondary and Tertiary formations (Buckland, 1820; Cuvier, 1813; Phillips, 1829–1836). First, it was believed that the Primitive rocks were covered by an average of at least two miles of Secondary and Tertiary strata, which showed slow deposition during successive periods of calm and catastrophe. Second, some strata had clearly formed from the violent destruction of older strata. Third, different strata contained different fossils; it was especially noted that strata containing terrestrial and fresh-water shells alternated with those containing marine shells, and that strata nearest the surface contained land animals mixed with marine creatures. Fourth, the lower strata generally showed a greater difference between fossil and living species, which suggested multiple extinctions, attributed to a series of revolutions over a long time. Fifth, evidence that faults and dislocations occurred after the induration of many strata implied a lapse of time between their formation and that of overlying strata. Finally, man was apparently only found fossilized in the most recent strata. Geologists concluded that the Earth was tens of thousands, if not millions, of years old and the relatively recent Noachian Flood affected only the rounded valleys and hills carved into consolidated strata and deposited only the superficial gravels and boulders (c.f., Buckland, 1820; Phillips, 1829).

By the end of the 1820s the major divisions of the geological record were well defined (Figure 1). The Primary rocks were the lowest and supposedly oldest, and were primarily igneous or metamorphic rocks devoid of fossils. The Secondary rocks were predominantly fossiliferous sedimentary strata. The Tertiary formations also contained many fossils, but they more closely resembled existing species. Atop all were the Alluvial deposits of gravel, loose sands, boulders and soil. More detailed local classifications were not always easily correlated.

A massive blow to catastrophism came between

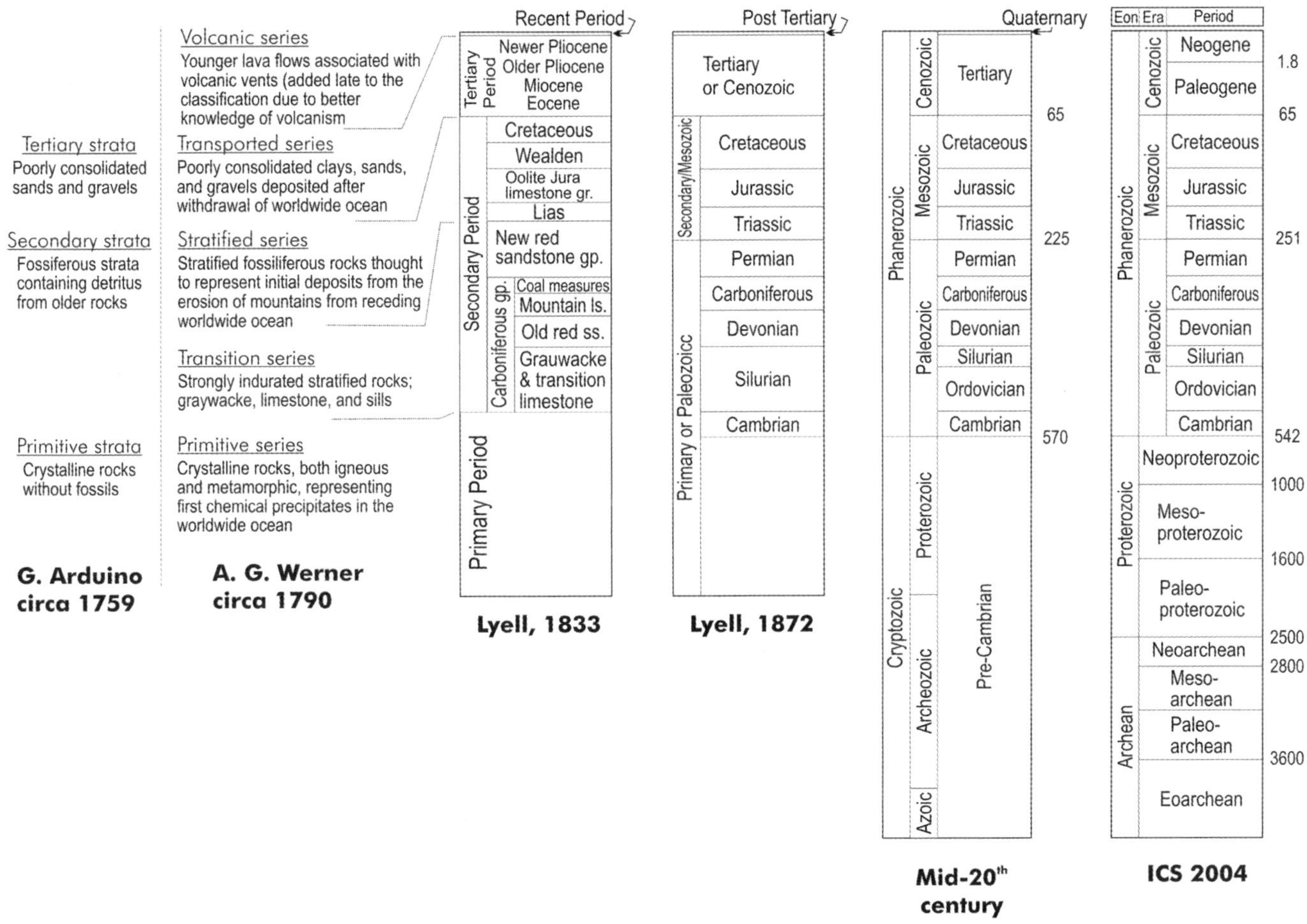

Figure 1. Chart showing development of geological timescale in England. Modified from Rudwick (1985) and Gradstein et al. (2004).

1830 and 1833, when Charles Lyell (1797–1875), a lawyer by training as well as a former student of Buckland, published his masterful three-volume work, *Principles of Geology*. Reviving the ideas of Hutton, Lyell's *Principles* set forth geological methods based on a radical uniformitarianism: only present-day erosional, sedimentary, volcanic and tectonic processes at present-day rates of intensity, frequency and magnitude should be used to interpret past geological activity.

Although the catastrophist theory had greatly reduced the geological significance of the Noachian Deluge and expanded Earth history well beyond the traditional biblical view, Lyell's work was a fatal blow to the Flood, explaining the whole rock record by slow gradual processes, thereby reducing the Flood to a geological non-event. His theory also expanded the length of earth history even more than catastrophists had. Lyell saw himself as "the spiritual saviour of geology, freeing the science from the old dispensation of Moses" (Porter, 1976, p. 91). Catastrophism did not die out immediately, although by the late 1830s few old-earth catastrophists in Britain, America, or Europe believed in a geologically significant Noachian Deluge.

Uniformitarian and catastrophist old-earth theories of the early 19th century did not go uncontested (Figure 2). During the 1820s to 1840s a group of Christian writers, who collectively became known as "scriptural geologists," published their biblical, geological and philosophical objections in pamphlets, journal articles, and (short or quite long) books. Some of these men were clergymen, some were not. Some were scientists and others not. And as was common at this time, some of the scriptural geologists were both clergymen and scientists. Some were geologically competent by the standards of their day (both by reading and geological field work) to critique the old-earth theories. Many of their biblical and geological objections are similar to arguments used by young-earth geologists today. The primary sources of this period show that the writings of the most geologically competent scriptural geologists were ignored and their arguments never refuted (Mortenson, 2004). While their biblical arguments were not identical and they disagreed on minor details, the scriptural geologists all believed (in harmony with most of the church for 1,800 years) that God had created the world in six literal days about 6,000 years ago and that most of the geological record was the result of the global, year-long, catastrophic Noachian Flood.

Catastrophist View (e.g., Cuvier, Smith)

[SB]--------- [C]--------- [C]--------- [C]--------- [C]--------- [C]-------- [F]----------- [P]

Uniformitarian View (e.g., Hutton, Lyell, Buffon, Werner)

[SB?] -- **[P]**

(up to millions of years)

Traditional Biblical View (Scriptural geologists)

[SCW] --------- **[F]**---------------------------- **[P]**

(circa 6000 years)

Figure 2. Timelines of earth history according to the three views of the early 19th century. SB = supernatural beginning; SCW = supernatural creation week of six literal days; C = catastrophic flood of regional, continental or global proportions; F = Noah's Flood; P = present.

Fine-tuning the Geological Column and its timescale

From the mid-1820s geology rapidly matured. Smith's stratigraphic methodology (using fossils to correlate the strata) was applied more widely by a growing body of geologists to produce more detailed descriptions and maps of the geological record. There was still debate over the nature and origin of granite, and Cuvier's widely-accepted interpretation of the Paris

Eon	Era	Epoch	Date	Basis for stratigraphic nomenclature
Phanerozoic	Cenozoic	Holocene	1885	Third Intl. Geological Congress renames Lyell's "recent"
		Pleistocene	1839	C. Lyell, eastern Sicily, based on % modern mollucsks
		Pliocene	1833	C. Lyell from mollusk studies in England, France, Italy
		Miocene	1833	C. Lyell from mollusk studies in England, France, Italy
		Oligocene	1854	H.E. von Beyrich, northern Germany and Belgium
		Eocene	1833	C. Lyell from mollusk studies in England, France, Italy
		Paleocene	1874	W.P. Schimper, based on rocks in the Paris Basin
	Mesozoic	Cretaceous	1822	Conybeare/Phillips(England) grouped chalk and lower strata, following Smith's earlier 4-fold division
		Jurassic	1839	L. von Buch formalized von Humbolt's Jura Limestone from Jura Mtns. in Switzerland.
		Triassic	1834	F.A. von Alberti grouped three distinct units into "Trias" in southern Germany
	Paleozoic	Permian	1841	Murchison from type section in Urals, near Perm, Russia
		Carboniferous	1822	Conybeare and Phillips formalized name for coal-bearing strata in northern England
		Devonian	1839	Sedgwick & Murchison, type section near Devon, England
		Silurian	1839	R.I. Murchison from type section in Wales
		Ordovician	1879	Lapworth identified new epoch to resolve Cambrian-Silurian controversy (Murchison vs. Sedgwick)
		Cambrian	1835	A. Sedgwick from section near Cambria (northern Wales)
Precambrian				Informal name for everything beneath Cambrian, formal divisions (i.e., Archean) are based on absolute dating

Figure 3. Nomenclature for the column developed during the mid 1800s.

basin was being challenged. By the early 1830s, all the main elements of stratigraphic geology were established, and maps and journal articles became more technical as geology made the transition from amateur avocation to professional vocation. The 1830s and 1840s saw much debate about the classification of the lowest fossiliferous formations (the Cambrian to Devonian), and the glacial theory emerged to explain geological features which the earlier catastrophists had attributed to the Flood. By the mid-1850s all the main time periods had been identified and the nomenclature standardized (Figure 3). However, none of these developments added any fundamentally new reasons for believing in a very old Earth, and all of this fine-tuning was based on principles established by Cuvier, Smith, Lyell and others.

In the last half of the 19th century, estimates of deep time increased and became more precisely quantified. In 1863 the great English physicist Lord Kelvin (1824–1907) used thermodynamics related to the internal heat of the Earth to argue for a "young" age of the Earth of only about 98 million years, much too short for Darwin's theory (which he rejected). But his view did not hold sway for long. Radioactive isotopes were discovered in 1896 by the French physicist Henri Becquerel (1852–1908). Heat from their decay negated Kelvin's conclusion, supporting the geologists' calls for hundreds of millions of years. Between 1903 and 1906, the famous New Zealand physicist Ernest Rutherford (1871–1937) determined that isotopes could be used to date rocks. By the 1930s, through the efforts of Arthur Holmes, the age of the earth had expanded to about 2 billion years, and has since risen to ~4.6 billion years (Robb et al., 2004). Modern additions to geology (e.g., plate tectonics) are merely refinements of the uniformitarian-naturalistic worldview.

In the last part of the 20th century, two unexpected ideas emerged. In 1961, the young-earth creationist movement was launched by the publication of *The Genesis Flood* by John Whitcomb and Henry Morris, who had been influenced by a few critics of old-earth geology in the first half of the 20th century. Since then, creationists have been hammering out an increasingly sophisticated model of origins (Morris, 1993). In the 1970s, some secular geologists also began to question uniformitarian assumptions. Returning to ideas of the early 19th century catastrophists, these "neo-catastrophists" reinterpreted many formations to be the result of rapid, violent catastrophic floods (Ager, 1973; 1993). It remains to be seen if the geological establishment will continue to ignore the work of modern young-earth geologists, as they did the writings of geologically competent scriptural geologists.

Key Issues in Old-Earth Theory

To understand the historical development of the old-earth theory and the extent of the validity of the conventional geological column, it is important to carefully consider several key issues.

Naturalism

Philosophical assumptions drove the development of the old-earth theories in the early 1800s. Two key assumptions were: (1) everything in the physical universe can and indeed must be explained by time, chance and the laws of nature working on matter; and (2) natural physical processes have always acted in the same manner, rate and intensity as we see operating today. These assumptions form the basis of uniformitarian methodological naturalism, which took control of modern science in the early 19th century, decades before Darwin published *Origin of Species* in 1859. Although many scientists today allow large-scale catastrophes, uniformitarian thinking is still endemic and naturalism is king. So, the heart of the debates about the age of the Earth and about how to correctly interpret the geological record is a massive worldview conflict.

Many old-earth pioneers expressed their worldview clearly. For example, Buffon wrote:

> In order to judge what has happened, or even what will happen, one need only examine what is happening.... Events which occur every day, movements which succeed each other and repeat themselves without interruption, constant and constantly reiterated operations, these are our causes and our reasons (cited in Gillispie, 1970–1980, vol. 2, p. 578).

Hutton similarly wrote:

> The past history of our globe must be explained by what can be seen to be happening now.... No powers are to be employed that are not natural to the globe, no action to be admitted except those of which we know the principle (cited in Holmes, 1965, pp. 43–44).[12]

Obviously, such statements ruled out a priori God's supernatural creation of the world in six days and the supernatural, global, year-long, catastrophic Noachian Flood. Werner, Laplace, Smith, Lyell and other leading developers of old-earth thinking followed this reasoning, and sadly many Christian geologists (Britain's Buckland and Sedgwick; America's Benjamin Silliman and Edward Hitchcock) were infected with this kind of thinking without realizing it.

Conscious rejection of Scripture

This naturalistic (deistic or atheistic) way of thinking developed in the social context of an overtly Christian culture in Europe and it was the result of a conscious rejection of Scripture. This anti-biblical worldview was often deliberately hidden from view in early published works which played lip-service to God's existence. But unpublished writings from the same men are more straightforward.

Buffon correctly perceived that his old-earth theory would be unacceptable to the Catholic Church. So, although his unpublished manuscript estimate 3,000,000 years for the age of the Earth, his published book gives an age of 75,000 years (which still was not acceptable to Catholic theologians). Jacques Roger, a leading 20th century French historian of science, says that "Buffon was among the first to create an autonomous science, free of any theological

[12] Holmes does not cite his source. The second half of his quote is found in Hutton's *Theory of the Earth* (Edinburgh: William Creech, 1795), vol. 2, p. 547. I could not find the first half of the quote in vol. 1 or 2 or in Hutton's 1788 journal article with the same title.

influence" (O'Connor and Robertson, 2004). Of course, he did no such thing. Rather, he wanted to enslave science to his own unbiblical theology and to "free" it from the Christian framework that gave birth to modern science.

In his *Theory of the Earth*, Cuvier briefly mentioned Genesis, the Creation, the Deluge and God, but dismissed all earlier efforts (including Burnett, Woodward and Whiston—no mention of Steno) to make sense of the geological record in light of those two events. He himself made no attempt to correlate his theory with biblical history except to allude to the post-Flood biblical chronology as giving a reasonable date for the Flood. But he did not specifically refer to any passage and ignored Genesis 1–9 and Exodus 20:8–11.

Lyell explained in a lecture on King's College London in 1832 that:

> I have always been strongly impressed with the weight of an observation of an excellent writer and skillful geologist who said that 'for the sake of revelation as well as of science—of truth in every form—the physical part of Geological inquiry ought to be conducted as if the Scriptures were not in existence (Rudwick, 1976, p. 150).

Such reasoning might be permissible, if the Bible did not describe any events relevant to the formation of the rocks of the Earth (such as the Creation Week and the Flood). But since the Bible does speak of such events, Lyell's approach is like trying to write a history of ancient Rome by studying the surviving monuments, buildings, artwork, and coins, but intentionally ignoring the writings of reliable Roman historians living at that time!

Lyell revealed his animosity toward the Bible and his devious plan to undermine its teachings in private correspondence. He confided in an 11 August 1829 letter to his friend and fellow old-earth geologist, Roderick Murchison, just months before the publication of Lyell's first volume of *Principles of Geology* (1830):

> I trust I shall make my sketch of the progress of geology popular. Old [Rev. John] Fleming is frightened and thinks the age will not stand my anti-Mosaical conclusions and at least that the subject will for a time become unpopular and awkward for the clergy, but I am not afraid. I shall out with the whole but in as conciliatory a manner as possible (Quoted in Brooke, 1979, p. 45, brackets added).

And to George P. Scrope (fellow uniformitarian geologist and Member of Parliament), Lyell wrote on 14 June 1830:

> I am sure you may get into Q.R. [*Quarterly Review*] what will free the science [of geology] from Moses, for if treated seriously, the [church] party are quite prepared for it. A bishop, Buckland ascertained (we suppose [Bishop] Sumner), gave Ure a dressing in the *British Critic and Theological Review*. They see at last the mischief and scandal brought on them by Mosaic systems.... Probably there was a beginning—it is a metaphysical question, worthy of a theologian—probably there will be an end. Species, as you say, have begun and ended—but the analogy is faint and distant. Perhaps it is an analogy, but all I say is, there are, as Hutton said, 'no signs of a beginning, no prospect of an end'.... All I ask is, that at any given period of the past, don't stop inquiry when puzzled by refuge to a 'beginning,' which is all one with 'another state of nature,' as it appears to me. But there is no harm in your attacking me, provided you point out that it is the proof I deny, not the probability of a beginning.... I was afraid to point the moral, as much as you can do in the Q.R. about Moses. Perhaps I should have been tenderer about the Koran. Don't meddle much with that, if at all.

> If we don't irritate, which I fear that we may (though mere history), we shall carry all with us. If you don't triumph over them, but compliment the liberality and candour of the present age, the bishops and enlightened saints will join us in despising both the ancient and modern physico-theologians. It is just the time to strike, so rejoice that, sinner as you are, the Q.R. is open to you.
>
> P.S. …I conceived the idea five or six years ago [1824–25], that if ever the Mosaic geology could be set down without giving offence, it would be in an historical sketch, and you must abstract mine, in order to have as little to say as possible yourself. Let them feel it, and point the moral (Lyell, 1881, I:268–271, brackets added).

So we could add one more assumption to early 19th century science: the Bible has nothing relevant to say to the question of the age and history of the Earth.

But none of this is surprising when we consider the theological orientation of the men who were most influential in the development of the old-earth theory. Buffon was a deist or atheist, disguising the fact with occasional references to God (Gillispie, 1970–1980). Laplace was an open atheist (Brooke, 1991). Lamarck straddled the fence between deism and atheism (Brooke, 1991). Werner was a deist (Page, 1969) or possibly an atheist (Hallam, 1992) and hence "felt no need to harmonize his theory with the Bible" (Gillispie, 1970–1980). Historians have concluded the same about Hutton (Dean, 1975). William Smith was a vague sort of theist, but according to his nephew (a fellow geologist) he was most definitely not a Christian.[13] Cuvier was a nominal Lutheran, but recent research has shown that in practice he was an irreverent deist (Brooke, 1991). Lyell was probably a deist (or a Unitarian, which is essentially the same) [Russell, 1985]. Many of the other leading geologists of the 1820s and 1830s were likewise anti-Christian. These men were hardly unbiased, objective pursuers of truth, as they would have wanted their contemporaries to believe and as modern evolutionists and historians of science want us to view them.

Shells and the Dating of Strata

Since shells made up the vast majority of fossils, they had a great, if not singular, importance for old-earth geologists. For example, William Smith, the "Father of English Stratigraphy," based his depiction of the geological column primarily on shells, which constituted the great majority of the fossils he listed in works on the geological record.[14] But he admitted that he did not know much about shell creatures in terms of species classification:

> …on this principle [of arranging the strata by their fossils] I have ventured, without much knowledge of Conchology, and with weak aids in that science (Smith, 1817, p. vi, brackets added).

In 1828 Lyell worked out his interpretation of the Tertiary (on which the first and later editions of his *Principles of Geology* depended) solely on the basis of shells (Lyell, 1863) even though he did not start studying "conchology" until 1830 (Lyell, 1881). Buckland stated that fossil shells were "of vast importance in investigating the records of the changes that have occurred upon the surface of our globe" and that "in fact without these [organic remains], the proofs of the lapse of such long periods as Geology shows to have been occupied in the formation of the strata of the Earth, would have been comparatively few and

[13] Smith's own writings reveal this vague theism, as do comments by geologist John Phillips, Smith's nephew and geology student. See Phillips (1844, p. 25). It is safe to say that Smith was definitely not a committed Christian.

[14] Smith (1816; 1817, p. vi and "Geological Table" after page xi). This table is reproduced in Sheppard (1914–22, opposite page 137).

indecisive" (Buckland, 1836, pp. I:110–112, brackets added). Geologist James Smith (1838) stated that judging the age of a deposit purely on conchological considerations was a sound rule of geological reasoning. So these "index fossils" were of critical importance as evidence for the old-earth theories.[15]

Several scriptural geologists raised objections to this use of fossil shells in dating the strata, referring both to uncertainties in taxonomic classification of shells and to ambiguities about their geological distribution. But they were not the only objectors. In both the 1812 and the 1831 editions of his *Theory of the Earth*, Cuvier rejected the use of shells as a means of reconstructing Earth's history, because differences in fossil species in the strata may have been the result of slight changes in salinity or temperature of the water or some other accidental cause, and because testaceous animals were still too poorly known to confidently claim that some were extinct (Cuvier, 1813; 1834). From 1808 to 1813 Beudant (to whose work the scriptural geologist George Young referred) had experimentally shown that marine shell creatures could adjust to life in fresh water and similarly fresh-water shellfish could become accustomed to life in the sea if the change in salinity was gradual as in the brackish waters of river deltas (Beudant, 1816). The old-earth geologist Macculloch (1824) referred to this and other observations about fish and shell creatures when he cautioned geologists about the use of these fossils to distinguish fresh water geological formations from those of marine origin. Six years later, he said that the use of fossils to identify, correlate and date strata from different locations was "groundless" and "nearly, if not entirely, useless" (Macculloch, 1831, pp. I:422–428, 453). In 1819 Greenough (pp. 302–304), then president of the Geological Society and an old-earth advocate, concluded that Cuvier's theory of the Paris Basin was open to "insurmountable objections," one of which was the difficulty of confidently distinguishing fresh-water and marine shells.[16]

Charpentier (1825), one of the leading geologists in Europe, argued that only the relative position of strata could indicate the relative ages of the rocks, because knowledge of fossils and their distribution in the strata was not sufficiently precise to use them as an index for dating. Also, the conchologist William Wood decried the "extreme multiplication of the genera [of shell creatures], rather to increase than remove the difficulty of determining the species" (1825, p. iv, brackets added). In an article on *mollusca* in the *Edinburgh Encyclopaedia* (1830) the old-earth zoologist John Fleming remarked on persistent difficulties in classifying shell creatures into species, genera and even the correct orders. The next year, De la Beche (1831) expressed strong caution in using shells to date strata, because of the considerable errors and confusion in the catalogues of fossil shells.

In the five editions of his *Introduction to Geology* published and revised between 1815 and 1838, the respected old-earth geologist, Robert Bakewell, repeatedly expressed his conviction that many of his fellow geologists relied too much on shells in their interpretations of the rocks: both in identifying distant, non-contiguous formations and in distinguishing fresh-water from marine deposits. He deemed this unwise because of the limited knowledge of shell creatures and the continuing evidence of much erroneous classification, especially the multiplication

[15] Other old-earth geologists said the same: Taylor (1829, pp. 26–41); Phillips (1837, pp. I:77–78); and Mantell (1839, p. I:202).

[16] This criticism of Cuvier was inaccurate, because, as noted above, Cuvier himself cited reasons why shells were not reliable indices and so why he built his theory of the Earth totally on the basis of quadruped fossils.

[17] Bakewell (1828, third edition, pp. 44–45) in this edition quotes from his 1815 second edition without giving the page numbers therein. See also 1833 (fourth edition, pp. iv–v, 42–43, 565) and 1838 (fifth edition, pp. 46–47, 397–404, 635).

of species and genera.[17] One reviewer of Bakewell's 1828 third edition apparently agreed with him about the dangers in applying conchological knowledge to stratigraphy (T. anonymous reviewer, 1829).

One of the most important challenges to the old-earth geologists' use of *mollusca* to identify and correlate rock layers was the work of John Gray (1800–1875), a leading conchologist at the British Museum. In 1833 he recorded the many difficulties and errors that had been made in classifying shell creatures based on the shells' features, which often resulted in the creation of many different species and genus names for what in reality was a single species. He cited examples where the shell nucleus (its embryonic form in the egg) of some large species had been mistaken for the full-grown shells of another species. He also noted that the nuclei of many shells of different genera do not have the same characters as their parent shells. He gave examples of shells that had swirls in opposite directions and so were classified as different species, when in fact they were the same. He showed that many species possess shells that are very regular when they are young, but whose adult forms are often irregular due to the substrate to which the animal attaches. Then there were the curious anomalies of shells that changed the direction of their last whorls as they approached maturity and sometimes even reverse the position of the mouth. and he noted that fractured shells had sometimes been named as different species.

Gray remarked: "The shape of attached shells depends greatly on the form of the bodies to which they are applied; and this is a circumstance that has been generally overlooked by conchologists" (1833, p. 781). He noted that authors had multiplied species on this basis. Gray demonstrated that the water's turbulence, the shell's exposure to light, temperature, and food would profoundly affect the thickness, roughness, color or size of the shells, leading many conchologists to multiply species. The article pointed out many other errors of conchologists (even leading experts) in the identification and classification of shell creatures at the species, genus, and family levels.

Two years later, in 1835, when the old-earth theory was more firmly established in geology (and Lyell's uniformitarianism was rapidly bringing catastrophists to the point of recantation), Gray published further results of his own observations and experiments which he explicitly applied to geology. He showed that some shells which appear to belong to the same genus are inhabited by very different animals and that some species of mollusks live in very different situations from the majority of species in the same genus, or they have the ability to maintain their existence in several different situations. This made it difficult to accurately discriminate different species and genera of living shell creatures and the problem was "much increased" in the case of fossil shells, especially when we have no strictly analogous living shells. For these reasons he stated that geologists had built their theories on much fallacious information about the species and genera of testaceous mollusks and he seriously called into question the propriety of using shells to distinguish and date strata.[18]

So, the scriptural geologists were raising a serious objection against old-earth theories when they contested the use of shells to date the rocks and work out geological history.

Shells as Index Fossils Today

But shell creatures are still very important to the dating of rock layers and the defence of the evolutionary view of the geological record. The respected and detailed reference work *Index Fossils of North America* (Shimer and Shrock, 1944) lists all the

[18] It is noteworthy that both Mantell (1839, second edition, p. I:202) and Phillips (1837, p. I:78) cited Gray's 1833 article, but not his 1835 article. William Buckland referred to neither article in the various discussions of shells in his *Bridgewater Treatise* (1836).

index fossils used by geologists to identify and classify rocks in the 1940s. All the fossils found in that text are invertebrates, such as protozoa (single-celled creatures), porifera (sponges), coelenterates (corals, jellyfish, etc.), echinoderms, annelids (e.g., worms), bryozoans, arthropods (trilobites, insects, etc.) and some plants (mostly of microscopic size). The largest two groups of index fossils are the brachiopods (bivalve shells) and mollusks (various kinds of other shell creatures such as snails, nautiloids, etc). Fish, amphibians, reptiles, mammals and birds are not listed.

In a short layman's book on the age of the Earth published by the Institute of Geological Sciences in London (Thackray, 1980), we find a chart (p. 10) of the key index fossils and they are largely shell creatures (Figure 4). In commenting on the chart (p. 8), Thackray informs readers that graptolites and ammonites are the most useful index fossils and he illustrates with a chart on page 13 (Figure 5). Here the strata are classified and correlated by ammonites, which supposedly represent about 15 million years of depositional history. Note that all the pictured ammonites which have been given different species names and in most cases even different genus names. One can only wonder if these classifications would remain valid under the critical eye of Gray back in 1835.

But there continues to be problems with classification of shell creatures. Ammonites are important index fossils, found in a wide range of sizes and in abundance from the Devonian to the Upper Cretaceous. Milton (1992) discussed their stratigraphic use in an Early Jurassic formation exposed in a pit

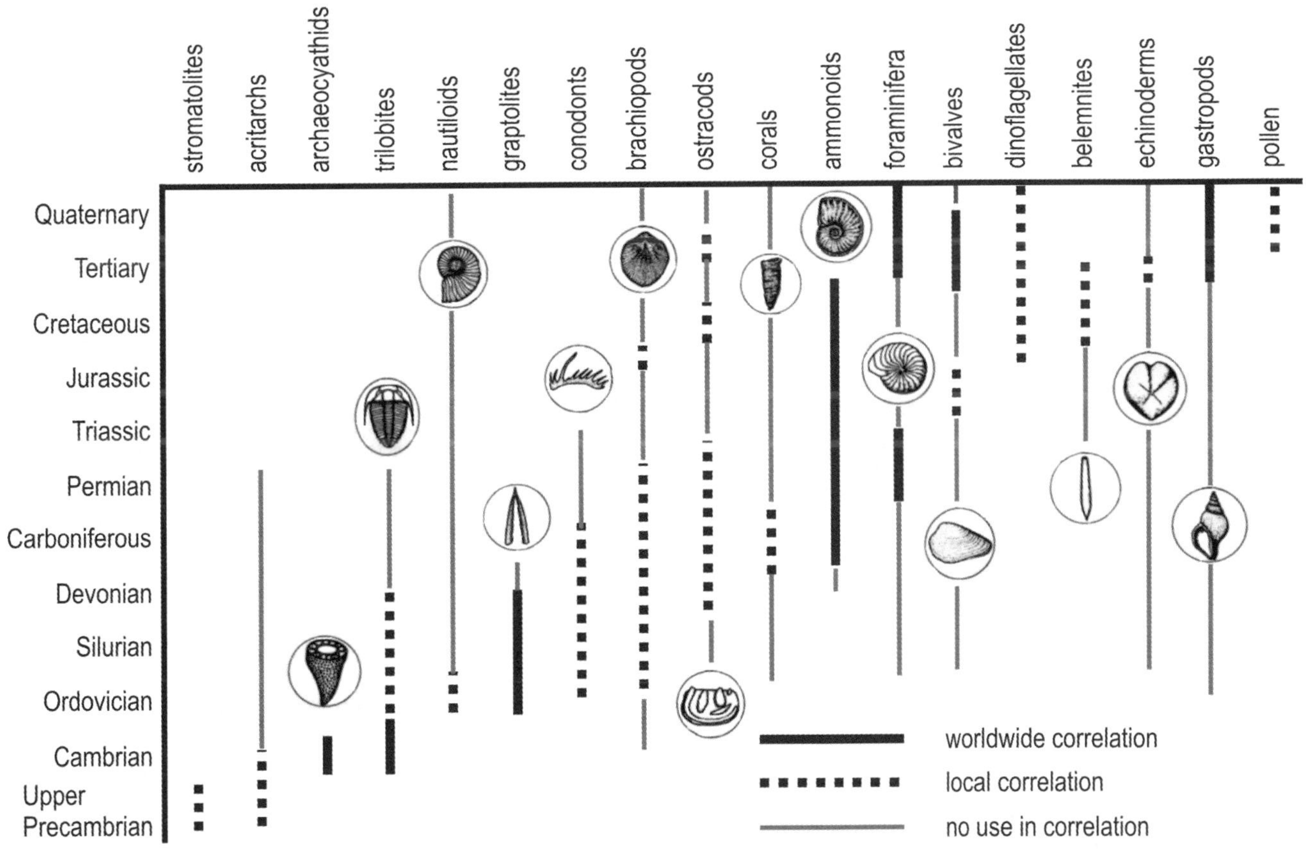

Figure 4. Ranges of index fossils, modified from Thackray (1980). Note that most are shell creatures.

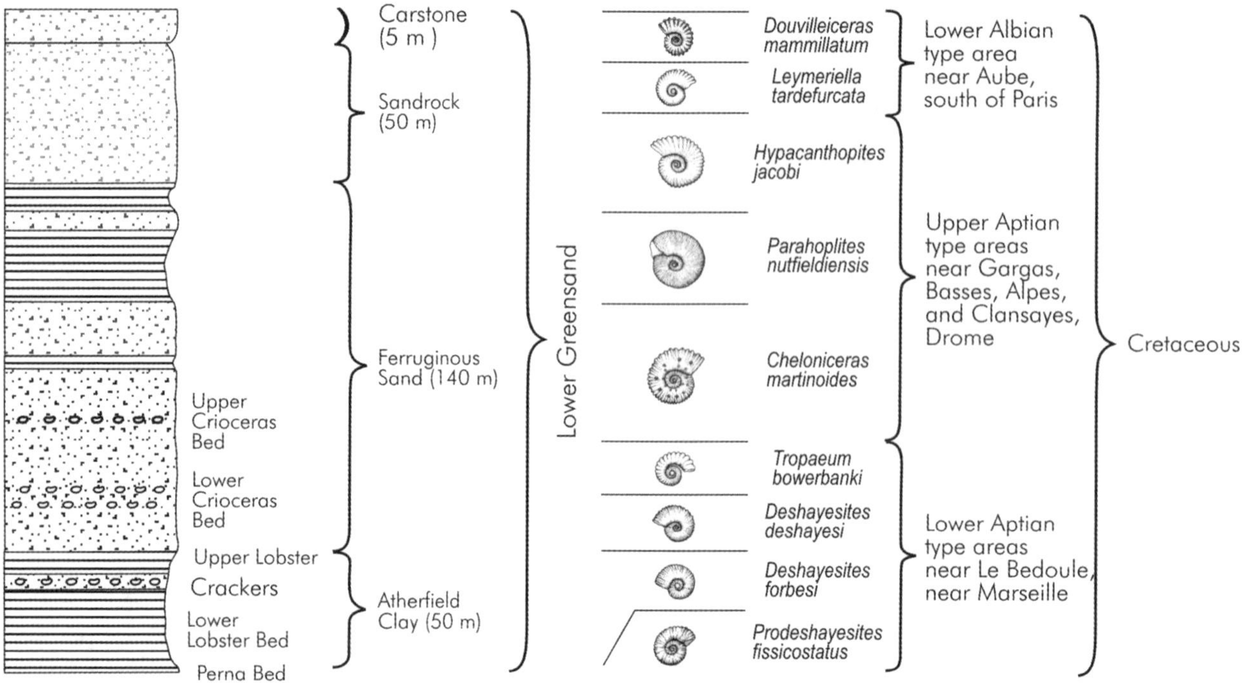

Figure 5. Thackray's (1980) representation of 15 million years of ammonite evolution.

in the village of Blockley, near Gloucester, England. Two ammonites dominate this formation: *Liparoceras* (fat, with two rows of knobs on the side) and *Aegoceras* (thin, with no knobs). A rare third form is said by evolutionists to be an intermediate between the first two, and has been labelled *Androgynoceras*, which resembles *Aegoceras* when it is young but looks like *Liparoceras* when it is old.

Milton noted that in 1870 an evolutionist named Hyatt arranged these ammonites in the evolutionary order (in ascending order—most ancient to most recent): *Aegoceras—Androgynoceras—Liparoceras*. But after careful examination in 1938 the famous palaeontologist Spath reversed the order: *Liparoceras—Androgynoceras—Aegoceras.* Then in 1963 the evolutionist Callomon re-examined them and rejected both explanations, insisting that *Androgynoceras* and *Liparoceras* are male and female of the same species. Following that example, Milton referred to volume L in the comprehensive 24-volume *Treatise on Invertebrate Paleontology* produced in the 1950s by the Geological Society of America. That volume illustrates and describes hundreds of ammonite species in minute detail. Under the heading "Iterative Evolution" the text states (as quoted by Milton, 1992, p. 108):

> The complexity of the modern classification, seen in the systematic parts of this Treatise, results mainly from acceptance in large measure of the theory of iterative evolution, although there are relatively few proved examples of its occurrence. We often feel sure that it has occurred and that we should be deceived if we accepted similarities at their face value (Haas, 1942 has brought together a few outstanding examples) but we can seldom demonstrate just what the 'iterative' relationships are.

Milton (1992, p. 108) adds that further on in volume L, the editor warns his readers, under the heading "Examples of ammonoid evolution," that:

> Waagen (1869) in a pioneer work attempted to demonstrate lineages, or lines of descent.... The chief obstacle to such studies is that a lineage is an oversimplified concept; it is *im-*

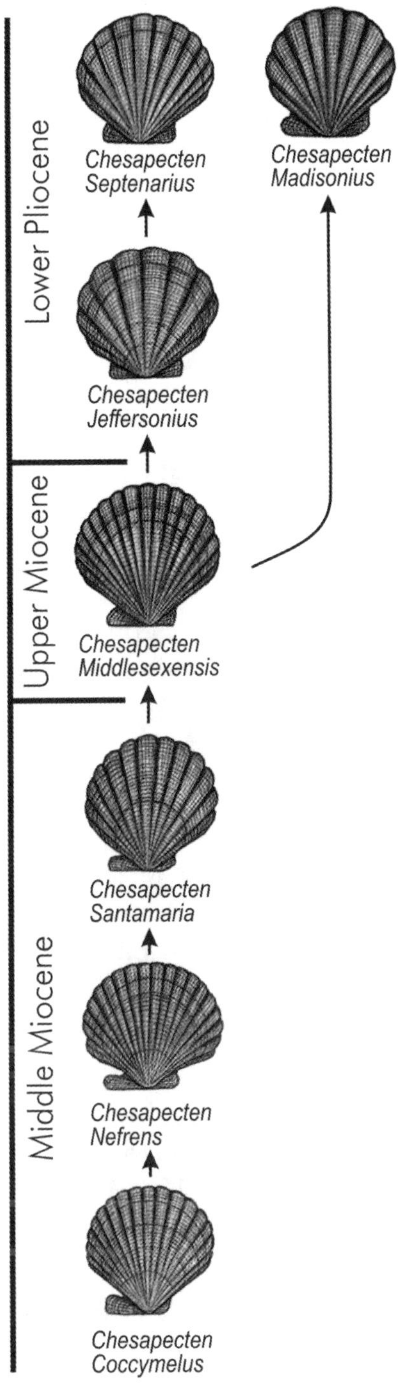

Figure 6. Supposed evolution of chesapecten shells over 10 million years. Modified from Levinton (1992).

possible to pick out a stratified succession of individuals which can with certainty be said to be genetically connected in the strict ancestor-descendant relationship (emphasis added).

Recent research by a Japanese team studying the 20 species of *Euhadra* snails has shown that changes in a single gene can change the direction of the shell spiral, thereby inhibiting, if not prohibiting, two snails with different shell directions to mate (Milius, 2003). The great evolutionary neo-catastrophist of the late 20th century, Derek Ager, discussed 10 quite different species of shell creatures and showed the significant changes in shell appearance of each species resulting from abrupt climate-induced changes in lake levels in sub-Saharan Africa (1993, p. 142). However, all his charts in the book showing the time-correlation of strata use only shell creatures (Ager, 1993, pp. 132, 133, 135, 141, 142, 148).

In discussing extinct Chesapecten fossils, Levinton (1992) used a diagram (Figure 6) to "show steady trends in evolutionary changes that persisted for more than 10 million years" (p. 90). But the differences are miniscule and mirror the development from juveniles to adults in modern scallops. Also, in an article on Chesapectens we are told:

> Paleontologists must name species based upon morphology, or shape. It is assumed that similar shape is a reflection of genetic relatedness as well. However, when one is designating a species, she must realize that there is inherent variation within species. Just as no two humans are exactly alike, no two clams are the same. This variation can be produced by simple genetic variation or environmental conditions. For example, some snails that live on rocky coasts may have thicker less ornamented shells than members of their own species that live in quieter water (Huntley).

Levinton's article does not even have accurate drawings of the shells (Figure 7). Just as Gray argued in 1835, geologists and paleontologists seem to still make too much of the variations seen in shell creatures.

In fact, modern experts on living shell creatures tell us that "Switching species from one genus to another

Figure 7. Photograph of *Chesapecten jeffersonius.* From http://en.wikipedia.org/wiki/Chesapecten_Jeffersonius. Contrast the front and back views of this shell with Levinton's (1992) second drawing from the top in Figure 6.

goes on all of the time because our knowledge of mollusks, indeed of all animals, is changing constantly" (Abbott and Dance, 1998, p. 5). The problems seem to be getting worse with regard to the fossil record. Sohn (2003) reports on the development of the Paleobiology Database, which was founded in 1998 with the goal of being an online repository of information about every fossil ever dug up. By August 2003 it had entered 340,000 specimens (from plants to whales to insects to dinosaurs to sea urchins) from 30,000 fossil collections and 10–15 scientists worldwide are adding to the database everyday. Sohn (2003, p. 33) remarked:

> Alroy has also used the database to reassess the accuracy of species names. His findings suggest that irregularities in classification inflate the overall number of species in the fossil record by between 32 and 44 percent. Single species often end up with several names, he says, due to misidentification and poor communication between taxonomists in different countries.

All of this gives us further reason to be suspicious of the validity of the geological column as a representation of the worldwide geological record that has been erected primarily on the basis of shell creatures.

The Absence of Fossils and the Non-Existence of Creatures

One more very important principle needs to be considered in the development of the geological column. That is the principle of judging the non-existence of a creature in history based on the absence of its fossil remains. Consider this discussion of index fossils in a standard geology textbook.

> No matter where on earth they are found individual species always occur in the same sequence relative to one another. By comparing fossils found in a layer of rock in one area with similar fossils in another area, we can correlate the two rock unites. More accurately, we can say that both rock units formed during the span of time that the species existed on the earth. Ideally, a geologist hopes to find an *index fossil*, a fossil from a very short-lived species known to exist during a specific period of geologic time. A single index fossil allows the geologist to correlate the rock in which it is found with all other rock layers in the world containing that fossil (Plummer and McGeary, 1993, p. 167, emphasis in the original).

This reasoning about index fossils is clearly based on the assumption that the absence of fossils in a certain layer of rock means that the creature did not exist at the time in which that layer was deposited. This has always been the way old-earth geologists have thought.

As Berry (1968, p. 59) summarizes in his book on the development of the geological timescale:

> William Smith provided geologists with a key by which the doors of past time might be unlocked: the principle of faunal succession led others to use specific fossils aggregates to delineate units in a timescale.

Smith's nephew and fellow geologist, John Phillips, put it this way in 1841:

> If instead of classifying the strata by mineral or chemical analogies, we resolve to employ the characters furnished by successive combinations of organic life which have appeared and vanished on the land and sea, we shall obtain an arrangement of remarkable simplicity, more

> precise in application, and yet less disagreeably harsh in definition, than that which has been so long followed. We shall have three great systems of organic life, characterizable and recognizable by the prevalence of particular species, genera, families, and even orders and classes of animals and plants, but yet exhibiting, clearly and unequivocally, those *transitions* from one system of life to another, which ought to occur in every natural sequence of affinities, dependent on and coincident with a continuous succession of physical changes, which affected the atmosphere, the land and the sea (quoted in Berry, 1968, p. 118, italics in original).

Based on the presence or absence of fossils Phillips was *assuming* that different kinds of creatures lived in different systems of life at different times. Of course, if the Flood occurred, as the Bible teaches, and it produced most of the geological record of sedimentary, fossiliferous rocks, then this old-earth thinking is very erroneous. The absence of fossils is no indication at all of the non-existence of a particular kind of creature.

Evolutionists use the idea that absence of fossils means nonexistence when it supports their theory and then ignore the principle when it does not support the theory. The late evolutionary giant Ernst Mayr (2001, pp. 63–64) reasoned:

> The fossil record of the earliest plants is very poor. Fossils of mosses generally considered the most primitive of the living land plants, have been found from the Devonian period, but surely they existed earlier but did not fossilize... The first vascular plants were found in the Silurian. The dominant plants in the Paleozoic era (particularly the Carboniferous) were lycopods, ferns, and seed ferns. The Mesozoic was dominated by gymnosperms... while the now dominant plants, the angiosperms, did not flourish until the Cretaceous... even though they originated in the Triassic. The entire enormous radiation of the flowering plants occurred since the middle of the Cretaceous, coevolving with a similar radiation of insects.

Mayr assumed the mosses existed earlier than we find them fossilized, because the evolution theory demands it. But he undoubtedly would have cried "foul" if creationists assumed that all plants and insects he mentions actually existed before their first appearance in the fossil record.

So-called "living fossils" are strong evidence of the fallacy of this assumption that absence of the fossils in a particular layer of rock means that the creature did not live at the same time as the fossilized creatures in that layer. Many examples can be given of creatures that evolutionists claimed for decades had long been extinct until living specimens were discovered which looked virtually identical to the fossil specimens. The Wollemi Pine was claimed by evolutionists to have gone extinct about 150 million years ago, because it was not found fossilized in rock layers assumed to be younger than that. Then it was discovered alive in 1994 and when it was compared to the fossil specimens it was seen to be virtually identical (Catchpoole, 2000a). The same could be said of the coelacanth fish and "Gladiator" insects (Catchpoole, 2000b).

Similarly, and related to our previous discussion on shells, slit shells (Family Pleurotomariidae, so called because of the slit part of the way around the shell between the last and penultimate spirals of the top-like shell) are called by evolutionists "the most primitive gastropods" (Hill, 2004, p. 8) because they are found in Upper Cambrian rocks (dated at 500 Ma). These deep-water shells were thought to be long extinct until they were discovered alive in 1855 (by Western scientists who didn't know that the Japanese had discovered them in 1772) and observed to be essentially identical to the oldest fossils (Hill, 1996).

Surely no one believes that these "living fossil" creatures actually went extinct for millions of years and then evolved back into existence to look just like their ancestors. Rather, they have been continuously alive since the first specimens were fossilized even though there was no fossil evidence of their existence in later deposited rock layers.

Thus, this fundamental assumption of old-earth geology is wrong. The absence of fossils in a particular rock layer does not mean that those creatures did not exist at the time of deposition of that layer. It only means that the creature was not buried in that layer or that it was but we haven't found it yet. Absence of fossil remains does not demand its nonexistence.

This erroneous thinking was especially applied to the antiquity of man relative to the Earth. A significant reason that the vast majority of geologists in the early 19th century believed that most of the geological record was deposited long before the creation of man was their conviction that no fossil human bones had been found except in recently formed deposits, and never with extinct animals.[19] Buckland (1836, p. I:103) said that "no conclusion is more fully established, than the important fact of the total absence of any vestiges of the human species throughout the entire series of geological formations." Clearly, even if that were true, it would not mean that man did not exist when the fossilized creatures did. The absence of evidence is not evidence of absence.

But many scriptural geologists argued that several fossilized human bones had been found in older rocks, and that this refuted the generally accepted opinion thereby militating against the old-earth theory.[20] They reasoned that if fossilized man was found in the same rock layer with any extinct creature, this fact would falsify the assumption that any other extinct creatures not found buried with man were necessarily in existence and became extinct before the creation of man.[21] At the time, these arguments by the scriptural geologists were ignored or dismissed by old-earth proponents on the contention that the cited fossil evidence had been misinterpreted either by the men who discovered the fossils or by the scriptural geologists who read their published reports.

However, not many years later Lyell (1863) and Phillips (1855) argued that several findings (cited by scriptural geologists) had indeed demonstrated that man had lived, died and was buried at the same time as some extinct animals. However, rather than argue that the rocks were deposited only a few thousands of years ago, Lyell and Phillips used the evidence to prove the great antiquity of man (beyond the biblical chronology)—even though the scriptural geologists had earlier used the same evidence to argue against the antiquity of the Earth. Lyell was not able to examine all of the physical evidence (which had been reported three decades earlier) because some of the sites had been destroyed by quarrying. But what he did see convinced him that the original investigators had provided "ample evidence" for their conclusions. He explained that the reason geologists back then (including himself) had not been willing to believe those conclusions was that the discoveries "contradict[ed] the general tenor of previous investigations" (p. 68, brackets added). The scriptural geologists, however, had contended that the reason for unbelief was that the findings contradicted the old-earth *theories*.

In addition, several scriptural geologists contended that the argument for the non-existence of man (or indeed any other creatures) in earlier times, based on the absence of fossil evidence, was philosophically

[19] c.f., Conybeare and Phillips (1822, p. lix); Lyell (1830–1833, pp. I:153–54).

[20] These included Fairholme, Murray, Young, Penn and Bugg. See Mortenson (2004).

[21] In other words, just because the dodo bird is now extinct (becoming so in the 18th century) does not mean that dodo birds became extinct before man was created.

unsound. They argued that since all contemporary creatures do not live in the same ecological habitat, it is unreasonable to expect them to be buried together. Also, geologists had only examined a very small portion of the Earth's strata. Furthermore, if during the Flood much of the antediluvian continents had been submerged to become post-diluvian ocean bottoms, most humans would have been buried out of the reach of geological investigation.

But some old-earth geologists also found the argument for non-existence of creatures based on the absence of fossil evidence to be problematic. Phillips (1837, p. I:96) said it led to erroneous conclusions about the history of birds. Smith (1838, p. 84) remarked that it would result in false inferences about the history of man in the British Isles. Lyell argued that erroneous conclusions about the history of fishes were produced by such reasoning, and in 1855 he provided a table documenting the previous 100-year history of the gradual discovery of different classes of fossil vertebrates in lower (i.e., older) formations than had been previously expected. He ended the discussion by saying:

> In conclusion, I shall simply express my own conviction that we are still on the mere threshold of our inquiries; and that, as in the last fifty years, so in the next half-century, we shall be called upon repeatedly to modify our first opinions respecting the range in time of the various classes of fossil Vertebrata. It would therefore be premature to generalize at the present on the non-existence, or even on the scarcity of Vertebrata, whether terrestrial or aquatic, at periods of high antiquity, such as the Silurian or Cambrian (Lyell, 1855, p. 463).[22]

[22] Lyell, of course, had his own agenda in saying this. At the time he was still very much opposed to the idea of progression (or evolution) in terms of plant and animal history, favoring instead a cyclical uniformity to life. See Gould (1987, pp. 132–142) and Ager (1993, p. xvii).

This is a significant revelation, given Buckland's confident assertion in 1836 (p. I:113):

> The deeper we descend into the strata of the Earth, the higher do we ascend into the archaeological history of past ages of creation. We find successive stages marked by varying forms of animal and vegetable life, and these generally differ more and more widely from existing species, as we go further downwards into the receptacles of the wreck of more ancient creations. When we discover a constant and regular assemblage of organic Remains, commencing with one series of strata, and ending with another, which contains a different assemblage, we have herein the surest grounds whereon to establish those divisions which are called geological formations, and we find many such Divisions succeeding one another, when we investigate the mineral deposits on the surface of the Earth.

Lyell's observation seems to be confirmed on a regular basis in our own time. For example, recently paleontologists digging in China and Mongolia have unearthed thousands of well preserved salamanders in rocks supposedly 165 million years old, although previously the earliest they had been found was in rocks estimated to be 65 million years. These new fossils resemble salamanders found today in North America and Asia (Gao and Shubin, 2003). Many other "living fossils" virtually indistinguishable from their supposedly ancient ancestors could be cited.

Conclusion

Many of the pioneers in the development of the science of geology worked within a biblical framework. They saw in the geological record a testimony to the truth of Genesis regarding Creation and the Flood. But in the late 18th and early 19th centuries, several false assumptions began to control geological thought.

Uniformitarian methodological naturalism divorced geology from the Bible and excluded God's mighty acts of creation and judgment from history. This rejection of Scripture's testimony to geologically significant events in history was conscious and intentional, and driven by anti-biblical worldviews (e.g., deism and atheism). In spite of their great ignorance of living and fossil creatures and the fossil record as a whole, leading geologists based stratal successions on questionable variations in shell creatures. They also assumed that the absence of fossil remains in particular strata was unambiguous evidence for their historical absence worldwide. Errors in logic and empirical errors were both built into their history of the Earth. Unfortunately, the old-earth proponents at that time never really engaged the arguments of their well-informed critics (scriptural geologists). That continues today—old-earth proponents commonly ignore young-earth creationists and their arguments.

The correct interpretation of the geologic record can only be obtained by using biblical assumptions gleaned especially from the history of Genesis 1–11. Young-earth creationists must help each other to see when they are unconsciously reasoning with the anti-biblical assumptions developed by the scientific establishment over the past 200 years. Otherwise, the rocks will continue to be a confusing mystery.

References

Abbott, C. and P. Dance. 1998. *Compendium of seashells.* Odyssey Publishers, El Cajon, CA.

Ager, D. 1973. *The nature of the stratigraphical record.* MacMillan Press, London.

———. 1993. *The new catastrophism.* Cambridge University Press, Cambridge.

Bakewell, R. 1815. *An introduction to geology,* second edition. J. Hardy, London.

———. 1828. Third edition, Longman & Co, London.

———. 1833. Fourth edition, Longman & Co, London.

———. 1838. Fifth edition, Longman & Co, London.

Berry, W.B.N. 1968. *Growth of a prehistoric timescale.* W.H. Freeman, San Francisco, CA.

Beudant, F.S. 1816. Extract from a memoir read to the Institute on the 13th of May 1816 on the possibility of making the Molluscae of fresh water live in salt water, and vice versa. *Philosophical Magazine* XLVIII(22):223–227.

Brooke, J.H. 1979. The natural theology of the geologists: some theological strata. In Jordanova, L.J. and R.S. Porter (editors). *Images of the Earth.* British Society for the History of Science, Monograph 1.

———. 1991. *Science and religion.* Cambridge University Press, Cambridge.

Buckland, W. 1820. *Vindiciae Geologicae.* University Press, Oxford.

———. 1836. *Bridgewater Treatise.* 2 vol. John Murray, London. [Note: The full title of this two-volume work was *On the power, wisdom and goodness of God as manifested in the Creation: geology and mineralogy considered with reference to natural theology*, but as one of the eight *Bridgewater Treatises* it is often referred to with the latter title.]

Catchpoole, D. 2000a. Another 'living fossil' tree. http://www.answersingenesis.org/docs2/4416livingfossil_tree12-25-2000.asp.

———. 2000b. 'Living fossils' enigma. *Creation Ex Nihilo* 22(2):56. Also available at: http://www.answersingenesis.org/creation/v22/i2/living_fossil.asp (January 2006).

de Charpentier, J. 1825. On fossil organic remains as a means of distinguishing rock-formations. *Edinburgh Philosophical Journal* XII(24):320–321.

Coleman, W. 1973. Cuvier and Evolution. In Russell, C.A. (editor). *Science and religious belief.* Hodder and Stoughton, Sevenoaks, Kent, UK, pp. 229–234. Reprinted from Coleman, W. 1964. *Georges Cuvier, zoologist.* pp. 172–175. Harvard University Press, Cambridge, MA.

Conybeare, W. and W. Phillips. 1822. *Outlines of the geology of England and Wales.* Williams Phillips, London.

Cuvier, G. 1813. *Essay on the theory of the Earth.* Robert Kerr, translator. William Blackwood, Edinburgh.

———. 1834. *Researches on Fossil Bones.* 4 volumes. London.

Buffon, G. *Oeuvres Philosophiques,* p. 56A, quoted in Gillispie, C.C. (editor). 1970–1980. Buffon, Georges-Louis Leclerc, Comte de. *Dictionary of Scientific Biography.* volume 2, p. 578. Scribner, New York.

De la Beche, H. 1831. *A Geological Manual.* Treuttel and Würtz, London.

Dean, D.R. 1975. James Hutton on religion and geology:

the unpublished preface to his *Theory of the Earth* (1788). *Annals of Science* 32:187–193.

Fleming, J. 1830. Mollusca. *Edinburgh Encyclopaedia* XIV.

Gao, K. and N.H. Shubin. 2003. Earliest known crown-group salamanders. *Nature* 422:424–428.

Gillispie, C.C. 1951. *Genesis and Geology*. Harper and Brothers, New York.

———. (editor). 1970–1980. Buffon, Georges-Louis LeClerc, Comte de. *Dictionary of Scientific Biography,* volume 2, pp. 576–581. Scribner, New York.

———. 1970–1980. Cuvier, Georges. *Dictionary of Scientific Biography,* volume 3, pp. 521–528. Scribner, New York.

———. 1970–1980. Werner, Abraham. *Dictionary of Scientific Biography,* volume 14, pp. 259–260. Scribner, New York.

Gould, S.J. 1987. *Time's Arrow, Time's Cycle*. Harvard University Press, Harvard.

Gray, J.E. 1833. Some observations on the economy of molluscous animals, and on the structure of their shells. *Philosophical Transactions.* CXXIII, Part 2:771–819.

———. 1835. Remarks on the difficulty of distinguishing certain genera of testaceous mollusca by their shells alone, and on the anomalies in regard to habitation observed in certain species. *Philosophical Transactions.* Part 2:301–310. [Note: A one-page summary of this appeared under the same title in *Philosophical Magazine*, 3rd Serial, volume VII(39):210].

Greenough, G. 1819. *A critical examination of the first principles of geology*. London.

Hahn, R. 1986. Laplace and the mechanistic universe. In Lindberg, D.C. and R.L. Numbers (editors) *God and nature.* University of California Press, Berkeley.

Hallam, A. 1992. *Great geological controversies,* reprint of second edition. Oxford University Press, Oxford.

Hill, L. 2004. *The world's most beautiful seashells.* World Publications, Tampa, FL. [see also Tabor, B.K. 1996. Vistas. Smithsonian Institute Report No. 83, http://www.si.edu/opa/insideresearch/9683/9683vist.htm].

Holmes, A. 1965. *Principles of physical geology,* second edition. Ronald Press, New York.

Huntley, J. Chesapecten Evolution. NSF GK-12 Graduate Fellows Program, University of North Carolina at Wilmington, http://www.uncw.edu/smec/gk_fellows/Documents/ChesapectenEvolutionRM.pdf. (25 January 2006).

Laudan, R. 1987. *From mineralogy to geology: the foundations of a science.* University of Chicago Press, Chicago.

Levinton, J.S. 1992. The big bang of animal evolution. *Scientific American* (Nov.).

Lyell, C. 1830–1833. *Principles of geology.* John Murray, London.

———. 4 May 1832. Lecture II at King's College London on 4 May 1832. Quoted in Rudwick, M.J.S. 1976. Charles Lyell Speaks in the Lecture Theatre. *The British Journal for the History of Science* IX(32):150.

———. 1855. *Manual of Elementary Geology*. London.

———. 1863. *The Antiquity of Man*. London.

———. 1881. Full letter. In Lyell, K. (Lyell's sister-in-law). *Life, letters and journals of Sir Charles Lyell, Bart.* I:268–271. John Murray, London.

Lyell, K.M. (editor) 1881. *Life, letters, and journals of Sir Charles Lyell, Bart.* John Murray, London.

Macculloch, J. 1824. Hints on the possibility of changing the residence of certain fishes from salt water to fresh. *Quarterly Journal of Science* XVII(34):209–231 (especially 230–231).

———. 1830. Organic Remains. In Brewster, D. (editor) *Edinburgh Encyclopaedia* XV:753–754.

———. 1831. *A system of geology*. 2 vol., London.

Mantell, G. 1839. *Wonders of geology,* second edition. 2 vol., London.

Mayr, E. 2001. *What evolution is.* Basic Books, New York.

Milius, S. 2003. Your spiral or mine? *Science News* 164(16):243–244.

Milton, R. 1992. *The facts of life: shattering the myth of Darwinism.* Fourth Estate, London.

Morris, H.M. 1993. *History of modern creationism.* Institute for Creation Research, Santee, CA.

Mortenson, T. 2004. *The great turning point.* Master Books, Green Forest, AR.

O'Connor, J.J. and E.F. Robertson. 2004. Georges Louis Leclerc Comte de Buffon, http://www-history.mcs.st-andrews.ac.uk/Mathematicians/Buffon.html (accessed January 2006).

Page, L.E. 1969. Diluvialism and its critics in Great Britain in the early nineteenth century. In Schneer, C.J. (editor). *Toward a history of geology.* M.I.T. Press, Cambridge, MA.

Phillips, J. 1829–1836. *Illustrations of the geology of Yorkshire.* 2 vol., York, England.

Phillips, J. 1837–39. *Treatise on geology*. 2 vol. Cabinet Cyclopaedia, London.

———. 1844. *Memoirs of William Smith.* John Murray, London.

———. 1855. *Manual of geology*. Quoted in Anonymous. 1857. *Voices from the rocks*, pp. 83–85. London.

Plummer, C.C. and D. McGeary. 1993. *Physical geology*, sixth edition. William C. Brown, Dubuque, IA.

Porter, R.S. 1976. Charles Lyell and the principles of the history of geology. *The British Journal for the History of Science* IX(32):91.

Robb, L.J., A.H. Knoll, K.A. Plumb, G.A. Shields, H. Strauss, and J. Veizer. 2004. The Precambrian: the Archean and Proterozoic eons. In Gradstein, F.M., J.G. Ogg, and A.G. Smith (editors). *A Geologic Timescale 2004*. Cambridge University Press, Cambridge, UK.

Rudwick, M.J.S. 1985. *The meaning of fossils: episodes in the history of palaeontology*, second edition. University of Chicago Press, Chicago.

Rupke, N.A. 1983. *The great chain of history*. Clarendon Press, Oxford.

Russell, C.A. 1985. *Cross-currents: interactions between science and faith*. IVPress, Leicester.

Sedgwick, A. 1831. Address to the Geological Society. *Philosophical Magazine* N.S. IX(52):314–315.

Sheppard, T. 1914–1922. *Proceedings of the Yorkshire Geological and Polytechnic Society* N.S. XIX(1914–1922): opposite page 137.

Shimer, H.W. and R.R. Shrock. 1944. *Index fossils of North America*. John Wiley, New York.

Smith, J. 1838. On the last changes in the relative levels of the land and sea in the British Islands. *Memoirs of the Wernerian Natural History Society.* III:84–85.

Smith, W. 1815. *A Memoir to the Map and Delineation of the Strata of England and Wales, with part of Scotland.* London.

———. 1816. *Strata identified by organized fossils.* W. Arding, London.

———. 1817. *Stratigraphical system of organized fossils.* E. Williams, London.

———. 1835. *Deductions from established facts in geology.* Scarborough, England.

Sohn, E. 2003. The fossils files. *New Scientist* 179(2409):33.

T. 1829. Anonymous review of Robert Bakewell's *An Introduction to geology*, third edition. In *Magazine of Natural History* 1(4):355–356.

Taylor, R.C. 1829. Geological arrangement of British fossil shells. *Magazine of Natural History* II(6):26–41.

Thackray, J. 1980. *The age of the Earth.* Institute of Geological Science, London.

Whewell, W. 1832. Anonymous review of Lyell's *Principles of geology*. *Quarterly Review* XLVII(93):126.

Winter, J.G. (translator). 1916. *Nicholaus Steno (1631–1686): The Prodromus of Nicholaus Steno's dissertation concerning a solid body enclosed by process of nature within a solid.* Reprinted in White, G.W. 1968. *Contributions to the history of geology, volume 4.* Hafner Publishing, New York.

Wood, W. 1825. *Index Testaceologicus; or a catalogue of Shells, British and Foreign.* London.

Young, D. 1988. *Christianity and the age of the earth.* Artisan Publishers. Thousand Oaks, CA.

Author

Dr. Terry Mortenson has degrees in math (BA), theology (MDiv), and history of geology (PhD). After serving for 26 years with Campus Crusade for Christ, primarily in Eastern Europe, he became a speaker, writer and researcher with Answers in Genesis in 2001. He and his wife, Margie, have eight children.

three

Toward a Diluvial Stratigraphy

John K. Reed
Peter Klevberg
Carl R. Froede Jr.

Debate about the nature and role of the geologic column points to the need for a diluvial overhaul of stratigraphy. After nearly two centuries of uniformitarian dominance, it is undergoing two revolutions. One is philosophical: the removal of its outmoded Lyellian straitjacket. The other is technological: the information age is bringing dramatic increases in our ability to image, measure, analyze, and visualize rock units. While the rapid retreat of the evolutionary establishment in the life sciences may be heartening to creationists, a concomitant rejection of Lyell's natural history lags behind. But the earth and life sciences share both a symbiotic interdependence and a mutual home in the worldview of naturalism; thus, the demise of one will force change in the other. Since uniformitarian geology preceded Darwin's synthesis and is more deeply entrenched, it is not unreasonable to expect it to outlive evolution.

What role will the Christian worldview have in that future? Uniformitarian scientists vehemently deny the recent impacts of creationism on stratigraphy, so they will probably belittle future influence, too. But their opinion about changes to stratigraphy is not as important as the changes themselves. Even though the numbers of creationists in the earth sciences remain small, the minority report has a tremendous advantage: a coherent philosophical framework for natural history that can fill the vacuum left by a collapsing naturalism. Scarce funding and man-hours can be offset by new technologies that leverage small resources into large returns.

Unfortunately, other factors retard diluvial progress. Ironically, the single largest is the ongoing inability to define a uniquely diluvial stratigraphy. That does not appear possible without resolution on the role of the uniformitarian geologic column. Context for that discussion requires background: the previous chapter provided a historical recap; this one will focus on the domain of stratigraphy itself. While not exhaustive, it will examine the discipline from a creationist perspective, outlining its fundamentals while hopefully clarifying needed reform in the light of an emerging biblical natural history.

Introduction

Stratigraphy is the study of the arrangement of earth materials (particularly stratiform rocks), especially as to geographic position and chronologic order of sequence (Bates and Jackson, 1984). Under the tutelage of uniformitarianism, the latter has been emphasized:

> It provides the means of piecing together a calendar so that world geologic records can be brought together into a coherent whole (Lemon, 1991, p. vii).

Diluvial stratigraphy will be distinct from its uniformitarian counterpart in several ways. At the heart of these differences is the central, but often ignored cornerstone of stratigraphy—*what is the relationship between science and history?* Naturalists would have us believe that the two are one and the same, that no knowledge can really be certain unless it is "scientific." Because Christianity is a religion that stresses God's actions in time, the science-as-history definition is unacceptable.

Other distinctives are obvious. Uniformitarian scientists emphasize the extended, monotonous march of history at a tempo no different from that observed today (with the occasional misstep of the local catastrophe); diluvialists stress the unique global catastrophe of the Genesis Flood. Uniformitarians measure deep time by the chronometers of evolution and isotopic decay; diluvialists default to the biblical narratives.

Despite these differences, there is common ground between uniformitarians and diluvialists with regard to stratigraphy because both share axioms native to Christianity.[1] Simply distinguishing between empirical and interpretive aspects of the discipline can identify much common ground. Both groups could probably accept a partial definition of stratigraphy as *the discipline of describing and classifying geologic data in three-dimensional space in terms of formations and relationships between formations with the goal of predicting what is (and was) in the subsurface via extrapolation (and interpolation) of observed data.* This definition provides a distinction between induction as a means to absolute truth (a fallacy) and induction as an indispensable part of the scientific method.

A Short History of Stratigraphy

Though practical application extends back over millennia, stratigraphy as a distinct discipline traces its roots to the scientific revolutions of the 17th century (Table I). The first recorded examination of the first principles of stratigraphy came from a Danish physician, Niels Stensen (1638–1686), also known as Nicholaus Steno, from his work in northern Italy. His initial interest in fossils led him to investigate the rocks which encased them. This work led to his three famous principles of stratigraphy:

- **Original horizontality.** Rock layers form in the horizontal position and deviations are caused by later disturbances.
- **Original continuity.** Rock layers extend laterally. Breaks within layers—such as valleys—postdate deposition.
- **Superposition.** Rock layers are arranged vertically, from the oldest at the bottom up to the youngest (age relationships are also determined by the principles of inclusions and crosscutting relationships).

[1] These Christian axioms dominated the culture of early exponents of naturalism to the extent that they may not even have been aware of their presuppositional theft. For more information, see Reed (2001).

System	Dates	Description
Diluvial	1550 - 1850	Emphasis on local geology; Deluge responsible for most rocks; Arduino introduces "primary," "secondary," and "tertiary"
Wenerian (neptunist)	1780 - 1840	Emphasis on lithostratigraphy; belief in successive worldwide lithologies
Cuvierist	1810 - 1850	Emphasis on biostratigraphy and unconformities; Deluges most recent in series of regional or global catastrophes
Lyellian (uniformitarian)	1830 - present	Emphasis on biostratigraphy; belief in series of worldwide ages characterized by fossil assemblages; continuity in natural processes
Neocatastrophist	1970 - present	Emphasis on unconformities; Lyellian history punctuated by regional or global catastrophes
Sequence Stratigraphy	1980 - present	Emphasis on unconformity bounded sequences, flooding surfaces, and physical boundaries; emphasis on eustatic cyclicity

Table I. Modern Schools of Stratigraphy

Steno's work was carried forward by Giovanne Arduino (1714–1795), a mining specialist working in northern Italy, who introduced the first chronology of the crust: the primary, secondary, tertiary, and quaternary. By the 18th century, there were two dominant European schools of geology, the Vulcanists (or Plutonists) and the Neptunists, who debated whether the rocks of Earth's crust were formed volcanically or hydraulically. One of the foremost Neptunists was the Prussian geologist, Abraham Gottlob Werner (1749–1817), who had a well-deserved reputation for his encyclopedic knowledge and his teaching. Werner was famous for his theoretical mastery of geology, but William Smith (1769–1839) pioneered a rigorous field emphasis. Smith, a surveyor and engineer who made careful observations of the lithologic and biostratigraphic features of common sedimentary layers, developed the first modern geological map in 1815, covering all of England and Wales—a map remarkably accurate even by today's standards (Winchester, 2001).

Early empirical work was soon overtaken by conceptual models. French anatomist Georges Cuvier (1768–1832), was an outstanding paleontologist closely associated with 18th century catastrophism. His position was quite distinct from biblical history in both its advocacy of multiple catastrophes—each resulting in the deposition of strata recording an extinction event—and in its elevation of forensic geology above biblical history. In many ways, it was similar to modern naturalistic neocatastrophism. More importantly, it typified the Enlightenment idea of a scientific history independent of Genesis.

Cuvier's catastrophism was itself forced to extinction by English uniformitarianism, which achieved dominance through the advocacy of Charles Lyell's (1797–1875) *Principles of Geology*. Lyell, a lawyer by training and an astute thinker, realized that a complete break with biblical history demanded the eradication of any catastrophism. Despite his reputation as a keen observer of geologic phenomena, Gould (1997, pp. 104–105) noted:

> Lyell's great treatise is not, as so often stated, a textbook summarizing all prevailing knowledge in a systematic way, but a passionate brief for a single, well-formed argument, hammered home relentlessly.

Despite its accumulated power and prestige, modern uniformitarianism is facing strong challenges from biblical and secular catastrophism. So far, it has survived by suborning each new development (e.g., sequence stratigraphy)[2] and dismissing diluvial thought out of hand with the worn-out adage of "religion vs. science." Though secular scientists are loath to admit it, creationism has played a large part in this breakdown,[3] penetrating even to Steno's axioms (Berthault, 1997). However, creationists have not yet achieved consensus on a coherent alternative. Given the current uncertain state, a review of stratigraphy, including an evaluation of its foundations from a creationist perspective, seems appropriate.

Underpinnings of Diluvial Stratigraphy

Though many aspects of modern stratigraphy are congenial to diluvial investigation, there are several important differences stemming from the worldviews of naturalism and Christianity. Most arise from their different views of science, history, and natural history. The most fundamental source of confusion in stratigraphy since its inception has been how to define natural history, specifically how to integrate science and history. Scientists operating within naturalism want to subsume much of history within science (Schlossberg, 1983) and sympathetic historians seem all too willing to concede (e.g., Collingwood, 1956). Though this reductionist approach has failed in its most explicit attempt (Marxism), it has met with greater success as a cloaked assumption. Christians have been slow to respond to this error and even today seem hesitant to address it head-on. However, any creationist stratigraphy *must* include a clear explanation of the interface between science, history (especially that of the Bible), natural history, and stratigraphy; anchored in and consistent with the biblical worldview.

The first step towards clarification is the realization that *any* interpretation of historical events lies ultimately outside science, because history concerns itself with unique past events while science addresses general principles by investigating the present.

> Though both history and science are investigative modes of inquiry that submit their conclusions to the test of experience (i.e., the data obtained by investigation), history by its method can answer questions that science cannot answer; and science by its method can answer questions that history cannot answer (Adler, 1993, p. 15).

Common ground is available; both history and science are built on theological and philosophical axioms (Clark, 1994). After a distinction is drawn between disciplines, creationists must draw another between the view of history inherent in their own worldview and that springing from naturalism.

One difference between the two frameworks arises from the notion of "prehistory" (Figure 1). Christianity recognizes that the beginnings of history were recorded by an observer—the traditional "data" of historical study. Naturalism believes that recorded history began quite late in Earth's timeline. Thus, the only record of prehistory is the rock record, accessible only by scientific study. Though it seems to make sense on the surface, this approach is classically circular because it allows the *conclusion* of deep time to function as an *axiom* of historiography. That sets the two worldviews in diverging directions. Naturalists cannot accept Christianity's insistence that all of time, including natural history, be constrained by the testimony of the Bible, nor can they accept the need for natural history to receive input from anything other than geology.

---

[2] Sequence stratigraphy has received mixed reviews from secular and creationist geologist (Bartlett, 1997; Berthault, 1997; Davison, 1995; Froede, 1994, 1997; 1998b; Holt, 1996; Klevberg, 2000b; Miall, 1986; 1992; Robinson, 1995; Thorne and Watts, 1984; R. Walker, 1990; T. Walker, 1996; Woodmorappe, 1996). See Appendix.

[3] Clark (1968, p. 29–36); Froede (1995); Mehlert (1986); Mortenson (2004); Reed (2001); Ritland (1981; 1982); Snelling (1992); Sunderland (1986); Taylor (1991); Whitcomb and Morris (1961, pp. 136–211); Woodmorappe (1980; 1981; 1983; 1996).

Naturalism	Christianity
Prehistory	No prehistory
Natural history is not constrained by revelation	Natural history must be constrained by revelation
Natural history must rely on rock record	Rock record can supply details not found in Bible
Study of rock record is science; therefore natural history is science	Control defaults to written record; therefore natural history is subset of history
Natural history is science	Natural history is a mixed question

Figure 1. Comparison of systematic consequences of the affirmation by Christianity and the denial by naturalism of "prehistory."

But natural history cannot be defined by science. Instead, history, science, and natural history must themselves be defined by philosophical reflection on revelatory truth. Some might question the relevance of the arcane mazes of philosophy to stratigraphy, but we must recall that science itself sprang from philosophy (it was formerly called "natural philosophy"), and is wholly dependent upon presuppositions of logic (proper reasoning) and metaphysics (the nature of reality). History also relies on philosophical conclusions about reality and knowledge. Thus, natural history—whether it is defined primarily as a subset of science or of history—must ultimately rest on philosophy. And unlike proponents of naturalism, creationists derive philosophical conclusions from biblical theology (Reed, 2000a).

Role of Science

Science certainly has a role in natural history; many aspects of stratigraphy are scientific. In the present context, stratigraphy is perhaps best pictured as the rope in a tug-of-war between two worldviews—naturalism and Christianity—with each side seeking scientific validation of its view of history. But neither side can win as long as they both expect the certainty of science to be seamlessly extended to historical interpretation. Clarity demands we address what science is not as well as what it is. Contrary to shallow definitions that equate it with naturalism, many[4] have pointed out that science depends on non-scientific presuppositions:

- A reality exists external to man.
- This external reality is ordered and causal.
- Sense perception is basically reliable.
- Man can comprehend nature through induction using rules of epistemology and ontology.
- Natural laws are unifiable and isotropic.
- Parsimony is desirable.

Although no precise definition of natural science is universally accepted (Moreland, 1989; Plantinga, 1997b; Schoepflin, 1982), the problem is not intractable, as evidenced in the similarity of various descriptions of the scientific method (e.g., Copi, 1982). Most include: (1) relevance. (2) testability, (3) compatibility with well-established hypotheses, (4) predictive or explanatory power, and (5) simplicity. Titus et al. (1979, pp. 201–208) divided the scientific method into two methods, logical (direct inferences, models, and paradigms derived from worldviews) and technical (means of observing or measuring the natural world). The latter generally include matters of agreement between creationists

[4] For those readers who wish to pursue this further, we recommend Ancil (1983; 1985), Bartz (1987; 1989; 1992), Bergman (1982), Brand (1985; 1996), J. Clark (1976), Frangos (1991), Hoffman (1993), Hull (1989a; 1989b), Johnson (1991; 1993; 1995), Kofahl (1986; 1989), Lillo (1987), Lumsden (1992), Malcolm (1997), McGhee (1987), Middelmann and Wilder-Smith (1980), Moreland (1989), Pearcey (1987), Pearcey and Johnson (2004), Plantinga (1990; 1997a; 1997b), Renard (1993), Riemen (1987), Schaeffer (1968a; 1968b; 1972; 1976), Sire (1988), Sproul et al. (1984), Verbrugge (1985), Øhrstrøm (1990). Some have attempted to provide a biblical approach in applying the philosophy of science to geology (Brand, 1974; Howe and Williams, 1994; Reed, 2001; Williams, 1992).

and evolutionists, the former do not. Both are subject to three tests for truth: (1) correspondence, (2) coherence, and (3) utility.

Today, the geologic establishment usually adds the caveat that science must default to natural explanation. They assume that since science is valid knowledge, there can be no truth outside natural explanation. Their approach flies in the face of logic (Clark, 1978; Pearcey and Johnson, 2004; Reed, 2001; Schaeffer, 1968a, 1968b, 1972, 1976; Sire, 1988), seeking truth, causality and order in all the wrong places. In building a successful stratigraphy, creationists must first escape this positivist trap.

Role of History

Just as science rests on axioms derived outside itself, so too does history. Naturalism approaches history in a manner that exhibits a degree of *internal* logic, but it cannot avoid fallacies of reductionism and circular reasoning. Metaphysical materialism leads inexorably to an epistemology of positivism. After all, if reality is matter, energy, and time interacting within physical laws, then science is the best hope for deciphering the universe. The rub comes with history; unique past events are not amenable to observation and experimentation. So naturalists seek resolution in uniformitarianism. If the past was no different than the present, then present observation (science) can be confidently extrapolated back in time (history). It was this *philosophical* meaning of uniformitarianism that opened the door in the 19th century for geology to replace biblical revelation, claiming both past and present for naturalism (Figure 2).

Uniformitarianism is seldom seen in this light because ever since naturalism's victory over biblical history, uniformitarianism is presented as no more than a forensic method, hiding its deeper axiomatic face. It is further confused by blurring uniformity and uniformitarianism; a classic "bait-and-switch" tactic—publicly presenting one definition while applying another (Reed, 1998). Creationists need to discern the philosophical wolf lurking beneath protestations of methodological innocence. Modern stratigraphy was nurtured by this semantic confusion and is today thoroughly permeated by it. That is why uniformitarians no longer see a lethal threat to their natural history from catastrophism; data do not penetrate that deeply into their paradigm.

However, this philosophical meaning of uniformitarianism is essential to naturalism in its continuing war against Christianity. First, natural history provides a supposedly scientific (and therefore certain) description of Earth's history—a sort of mythology for the antitheist. Second, it makes itself useful by offering through the geologic column a prepackaged framework for labeling and relating rock units as an essential task to economic exploration and development of crustal resources. Since physical correlation (outcrop, subsurface logging, or geophysical imaging) cannot be extended globally, the ability to pigeonhole rock units chronologically by their index fossils offers a resolution to field puzzles across lithology changes, faults, or sequence boundaries.

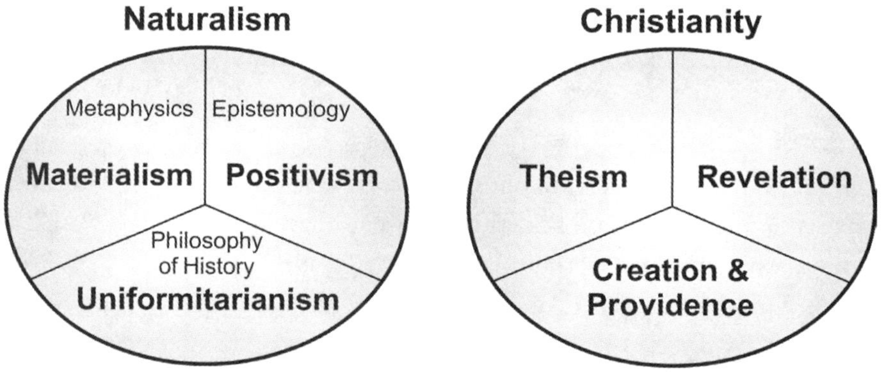

Figure 2. Comparison of naturalism and Christianity in their axioms of reality, knowledge, and history. From Reed (2001).

Distinctions between descriptive and genetic aspects of stratigraphy are seldom considered. However, the disguise is wearing thin; uniformitarians today have been reduced to appeals to authority and cries to defend "science" from the ancient evil of "religion."[5]

That weakness has not yet been fully exploited by Christians, perhaps because they are still too caught up in the modern euphoria of a scientific history. None of the early stratigraphers (many of whom were Christians) clearly addressed this issue, and their modern counterparts still fumble with unsatisfactory semantic distinctions between "operations science" and "origins science"—a distinction which avoids rather than addresses the question. Creationists can and should provide a better answer. At a minimum, this must include:

(1) The recognition that science and history are distinct disciplines[6] by their methods, goals, and resulting levels of confidence;
(2) The exposure and defeat of the philosophical meaning of uniformitarianism in favor of biblical testimony about the past;
(3) The refutation of the notion of "prehistory" external to the Bible; and
(4) The recognition that natural history is a means of filling in detail on the timeline, not extending it back into the indefinite past.

Biblical natural history and the role of stratigraphy

If Christians reject a positivist approach, is natural history even possible? The biblical worldview forces two distinctions between uniformitarian natural history and diluvial natural history. First, it must be a subset of history, not science, because it lies within the reach of recorded history. Second, it is a "mixed question"—one in which different disciplines interact to reach truth not achievable by the separate parts (Adler, 1965). Therefore, theology, philosophy, revelation, and science cooperate to draw conclusions about the natural events of the past not recorded in detail by human or divine witnesses. We agree with uniformitarians that there is historical content in the rock record, and thus stratigraphy is an important forensic tool. But the distinction between stratigraphy as a forensic tool of natural history and stratigraphy as the arbiter of natural history must be maintained. As an aside, the excesses of modern archeology could be corrected by the same remedy.

This arrangement has several noticeable benefits: (1) it lifts the awkward burden of historical interpretation from scientific methods; (2) it restores the confidence appropriate to science that is diminished when science makes claims about the past that are obviously fraught with uncertainty; and (3) it clearly distinguishes forensic observations from historical interpretation. Stratigraphy then becomes the tool of natural history rather than its synonym. We believe that this arrangement will work better in the long term for both stratigraphy and natural history, neither having to assume a role for which it is not fit.

How can creationists repair the decades of damage done by uniformitarianism? First, they must address, resolve, and continue to highlight ground rules of natural history (Reed, 2001, p. 19):

> The first step in such a critique is the recognition that historical analysis is a much larger and more complex question than is commonly presented in geologic interpretations. Key issues to be addressed prior to developing any model are: (1) the severely limited potential for human neutrality in historical analysis; (2) the proper domains and relationships

[5] The "war" between science and religion is another Enlightenment myth that is currently being debunked (Stark, 2003).

[6] The most cogent explanation of this relationship that we have seen is that of M.J. Adler in *The Conditions of Philosophy* (1965).

of the various areas of human knowledge; and (3) the criteria for establishing a critical framework within which competing models can be evaluated.

Uniformitarian natural history has failed to observe these differences, but has actively subsumed other fields of knowledge into science, scorning other methods as unscientific (Becker, 1932; Johnson, 1995; Plantinga, 1990; Schaeffer, 1968a, 1968b; Sire, 1988). This fallacy of reductionism seeks truth *within* the scientific enterprise instead of using the scientific method as one means among several to arrive at truth.

Second, a creationist natural history must redefine the forensic role of stratigraphy by setting boundaries, evaluating specific stratigraphic methods, and stripping out interpretive aspects embedded by years of uniformitarian distortion. After so many decades of uniformitarian rule, and despite protests of scientific purity, stratigraphic methods often contain unscientific elements, and methods that are scientific in theory are not always so applied (Figure 3). Recognition of the distinction between data and interpretation is vital if creationists are to reform stratigraphy or even to correctly decipher mainstream publications.

Third, creationists need to work towards a consensus on the geologic column. Naturalists uniformly accept the system of stratigraphic correlation by means of global synchronous time units. For creationists, it has not been so easy. The geologic column remains a topic of considerable controversy. Some have adopted the column, albeit with a compressed chronology, recommending that it be recognized as factual or substantially correct (Garner, 1996; Garton, 1996; Ritland, 1981, 1982; Robinson, 1995, 1996, 1997; Snelling et al., 1996). These and others (Mehlert, 1993a, p. 77) assert that the general biostratigraphic trends presented in the

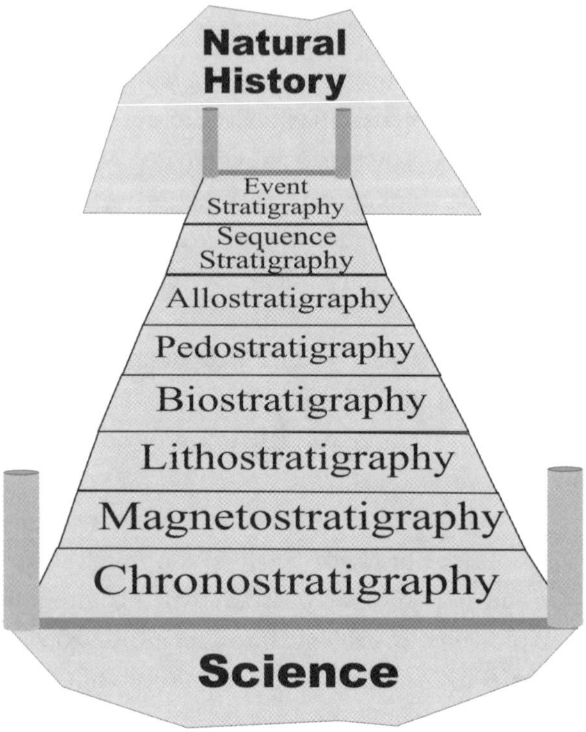

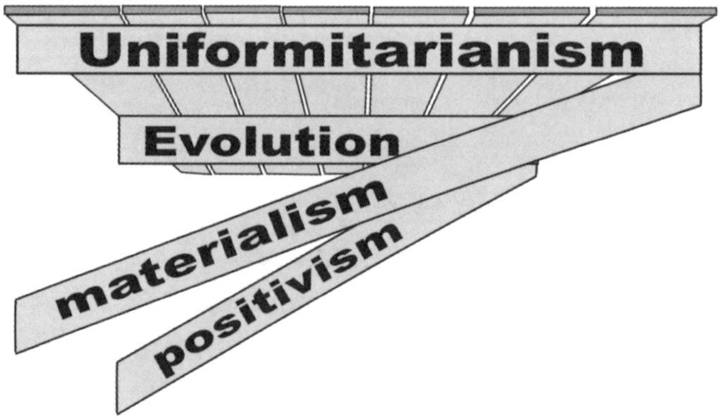

Figure 3. Uniformitarians claim that science can build a bridge to historical truth via stratigraphic methods (left), ignoring the perspective (right) showing the assumptions supporting their bridge.

geologic column are indisputable. Unfortunately, many creationists exhibit a surprising difficulty in differentiating between the enterprises of science and history—perhaps because they fear ceding an ability to apply scientific reasoning to the past, thus defaulting before the competition begins.

However, others have countered with inductive arguments from out-of-order fossils or formations (Lammerts, 1984a, 1984b, 1985a, 1985b, 1986a, 1986b, 1986c, 1987; Whitcomb and Morris, 1961), statistical analyses of fossil succession (Woodmorappe, 1981), statistical analyses of succession of geochronologic units (Woodmorappe, 1983; 1996; 1999a; Mehlert, 1993a; 1993b), inductive criticism of the column itself (Woodmorappe, 1981), and axiomatic criticisms of uniformitarian natural history (Reed, 2001). Others have challenged the unconfirmed use of worldwide unconformities (Morris, 1996, p. 55; Woodmorappe, 1978, p. 193), echoing secular workers (Vail et al., 1991, p. 621).

Creationists face a dilemma regarding the geologic column. On one side there is the danger of "reinventing the wheel" by not taking full advantage of existing knowledge. On the other is the danger of operating inconsistently with their parent biblical worldview. Despite many efforts, harmonization of biblical history and the geologic column has not resulted in any clear creationist stratigraphy. As Froede (1998, p. 2) stated: "Most, if not all, of these attempts to reconcile Scripture to the global evolutionary stratigraphic column have come at the expense of the biblical record." With many questions to answer and few resources at their disposal, creationists need to first reach coherence within their own worldview.

Once an integrated diluvial approach has been defined, creationists need to offset their numerical disadvantages with new technologies. Often this is difficult—for example, few individuals can afford the workstations, software, training and data to perform seismic analysis. Those tools will necessarily reside within research institutions, not yet funded to the extent necessary. However, budding creationists can consider employment in the oil or mineral industries to gain expertise in these methods. On the other hand, many technologies such as GIS software, satellite imagery, GPS, and many analysis and presentation products are affordable and should become standard in creationist research.

Empirical Stratigraphy and Natural History

Geologists expect to examine a given rock unit and answer questions about when and how it formed. However, the *science* of stratigraphy cannot answer these questions.

> Similarly, when scientists (such as geologists, paleontologists, and evolutionists) sometimes attempt to establish the spatial and temporal determinants of particular past events or to describe a particular sequence of such events, they cease to be engaged in scientific inquiry and become engaged in historical research, sometimes called natural history (Adler, 1993, p. 15).

Stratigraphy as a science must limit itself to empirical sources of knowledge: lithologic properties such as mineralogy, grain size, sorting, color, and chemistry; fossils; bedding characteristics of extent, thickness, grading, sequences, paleocurrent indicators, and nature and location of contacts; as well as derived properties of electrical resistivity, magnetic susceptibility, seismic impedance, porosity, transmissivity, and paleomagnetism. An empirical stratigraphy must meet typical scientific criteria of relevance, testability, compatibility with well-established hypotheses (scientific, not historical), predictive and explanatory power, and simplicity.

Non-empirical natural history uses the forensic conclusions from stratigraphy to draw conclusions about particular past events. Although forensic methods elevate certainty, they do not guarantee it. Many uniformitarian models are complex and coherent, but falsified because of assumptions of evolution and deep time. Christians face a different challenge: creationists know the outline of history from revelation, but lack the detail included in uniformitarian interpretations.

Therefore, stratigraphy as performed by creationists must: (1) be done in a context of its parent worldview, (2) respect boundaries between science and natural history, (3) develop rules for the application of science in historical investigations, and (4) follow methods appropriate to data type (e.g., difference between analysis of biblical text and seismic line). These all point inevitably back to the need to resolve the role of the geologic column in creationist stratigraphy. Is the column the outgrowth of empirical stratigraphy or is it the product of a biased natural history with significant non-empirical baggage?

As we seek those answers, it is well to remember that scientific stratigraphy may be practiced without complete resolution of these issues. Indeed, it can be performed ably by advocates of either worldview. Either may be a careful and relatively impartial observer, prudently differentiating observation from interpretation. Lack of personal objectivity may hinder practice, but the results should be the same within a narrow scope of research if proper methods are applied. It is the step from empirical stratigraphy to interpretive natural history that divides creationists from uniformitarians because of their radically varying presuppositions (Reed, 2001). As Davison (1995, p. 237) noted:

> A re-occurring problem in this exercise was the dependence on others' evolutionary-based interpretations—certainly not the ideal for trying to do a Flood-based interpretation…

Stratigraphic methods

Stratigraphy is not a single method, but a field of study that utilizes many. Unfortunately, it seems difficult to find geologists who are both familiar with the theoretical aspects of natural history and also expert in field studies. This may prove an advantage for creationists—those inclined to field careers are forced to become familiar with the theoretical background simply to defend their work within the context of their biblical worldview.

Present-day stratigraphy is codified in several places, most notably in the North American Stratigraphic Code (NASC) by the North American Commission on Stratigraphic Nomenclature (2005). The NASC defines terms and methods of stratigraphy in a clear and succinct manner, but also illustrates the uniformitarian inability to distinguish between science and natural history. On one hand, it pays lip service to the limits of observational science.

> The objective of a system of classification is to promote unambiguous communication in a manner not so restrictive as to inhibit scientific progress. To minimize ambiguity, a code must promote recognition of the distinction between observable features (reproducible data) and inferences or interpretations (p. 1556).

But in spite of these good intentions, further reading demonstrates that the committee cannot hide its assumptions about history. However, the NASC can be quite helpful in summarizing the present state of the discipline for any stratigrapher, and we strongly recommend it as a standard resource.

The NASC contains many other imperatives to distinguish between data and interpretation and maintain the time independence of descriptive methods, including lithostratigraphy, magnetostratigraphy, and biostratigraphy. However, creationists would disagree with its advocacy of isotopic dating (Vardiman et al., 2000; Gill, 1996; Helmick and Baumann, 1989; Jeletzky, 1978; Johansson, 1993; R. Johnson, 1993; Molén, 1991; Snelling, 1995, 1998; Woodmorappe, 1999b), and some would question the reality of plate tectonic rock assemblages often interpreted from biostratigraphy (Reed, 2000b).

Though their intention to separate empirical description from interpretation is often stated, they fail to follow their own advice many times by advocating stratigraphy as the home of historical interpretation.

> Stratigraphic procedures and principles, although developed initially to bring order to strata *and the events recorded therein*, are applicable to all earth materials, not solely to strata. They promote systematic and rigorous study of the composition, geometry, sequence, *history, and genesis* of rocks... (p. 1555–1556, emphasis added).

Chronology is considered a conclusion of stratigraphy, rather than a historical interpretation supported by it:

> Major objectives of stratigraphic classification are to provide a basis for systematic ordering of the *time* and space relations of rock bodies and *to establish a time framework for the discussion of geologic history* (p. 1558, emphasis added).

An emphasis is placed on chronostratigraphic correlation of rock units. This can only be done in the manner they prescribe by assuming deep time, uniformitarian conditions and the absence of a global flood.

> Correlation is a procedure for demonstrating correspondence between geographically separated parts of a geologic unit. The term is a general one having diverse meanings in different disciplines. *Demonstration of temporal correspondence is one of the most important objectives of stratigraphy* (p. 1560, emphasis added).

Their assumption of the historical veracity of evolution is sometimes subtle (note the unstated assertion that faunal content would be the primary basis for determining a difference in age for similar rock units).

> To facilitate the reconstruction of geologic history, some compositionally similar deposits in vertical sequence merit distinction as separate stratigraphic units because they are the products of different processes; others merit distinction because they are of demonstrably different ages (p. 1558).

Other times, it is not.

> ...the irreversibility of organic evolution makes it possible to partition enclosing strata temporally. Third, biologic remains provide important data for the reconstruction of ancient environments of deposition (p. 1558).

Neither evolution nor the temporal continuity of depositional environments is acceptable within a creationist natural history.

In spite of their shortcomings, the NASC does provide a description and a set of rules governing the application of various stratigraphic methods. Many are shown in Table II: minor variations and extra-scientific approaches have been omitted. Of these stratigraphic methods, six can be purely descriptive. These include lithostratigraphy, lithodemic stratigraphy, magnetostratigraphy, biostratigraphy, allostratigraphy, and seismic stratigraphy. However, they are seldom applied in purely descriptive fashion because uniformitarian scientists seem unable to separate their assumptions about history from their geologic investigations. The unfortunate result is that methods defined in empirical terms are often tainted by nonempirical concepts. Creationists cannot afford to make these same errors in delineating a new stratigraphy, and must beware of subjectivity compounded by observer bias and the difficulties of the task:

> In many local studies, facies descriptions may be so complex that they go beyond our interpretive abilities. The first way to simplify a complex scheme is to group facies perceived to be similar and/or genetically related... (Walker, 1990, p. 777).

Method	Description	Remarks
Lithostratigraphy	Correlation by rock type	Empirical in theory, subordinate to chronostratigraphy
Lithodemic Stratigraphy	Extension of lithostratigraphy to nonstratiform earth materials	Empirical in theory; subordinate to isotopic dating
Magnetostratigraphy	Correlation by rock magnetic properties	Empirical in theory; defined by geochronology
Biostratigraphy	Correlation by fossil assemblages; commonly by index fossils	Fossil succession and index fossils presuppose evolution
Pedostratigraphy	Correlation by paleosols	Requires recognition of ancient soil horizons/ weathering profiles
Chronostratigraphy	Correlation by assignment to geochronologic unit	Purely theoretical historical construct; rock units are material referents for specific geochronologic units
Polarity-Chronostratigraphy	Assignment to geochronologic unit based on magnetic polarity	Reference to chronostratigraphy imposes associated problems
Allostratigraphy	Correlation by lateral facies relationships in sedimentary basins	Empirical in theory; often in practice; limited applicability
Seismic Stratigraphy	Correlation by acoustic signature relative to surrounding waveforms	Potentially empirical, though purely geophysical
Sequence Stratigraphy	Correlation by stratigraphic sequences; controlled by eustasy	Ill-defined parameters; limited to marine margins; a priori assumptions
Transgressive-Regressive cycles	Correlation by reference to ocean margin migration	Limited to marine margins; subordinate to chronostratigraphy
Event Stratigraphy	Correlation by identification of single geologic event	Highly genetic; scale dependent; often combined with sequence stratigraphy

Table II. Stratigraphic Methods. Modified from Klevberg (2000a). References: Barlett (1997), NASC (2005), Seilacher (1991), Walker (1990).

Correlation—the goal of stratigraphy

A key goal of stratigraphic research is correlation. Correlation is the act of identifying various rock bodies over a laterally extensive area as pertaining to each other, for example, as part of the same rock unit, time period, time-rock unit, event, or sequence. The methods listed above all assume some application in this task. Correlation is most commonly done by direct physical properties when visible in outcrop, and by indirect electric, acoustic, magnetic, or radioactive measurements (i.e., geophysical logging and seismic methods) in the subsurface. Correlation, whether performed via direct observation or via indirect measurements, presents many pitfalls which require skill, training and vigilance to avoid.[7] Creationists can avoid some of those by rejecting those methods which rely on indemonstrable assumptions (i.e., evolution).

Establishing a global chronologic succession has been the major goal of stratigraphy since its inception, closely followed by the desire to relate a given stratigraphic unit to regional geologic processes. Correlation across time and space is the essence of the geologic column. All of the methods described in Table II help geologists define, describe, and correlate rock units. Thus, deficiencies in these methods are of great importance.

Both correlation and the methods described earlier are scale-dependent. Correlations made by dozens of well logs and cores within a 50-acre oil field differ greatly from megaregional reconstructions that interpolate over miles rather than meters. Large-scale correlation has been bolstered by sequence stratigraphy and its dependence on the more easily attained and larger-scale seismic data, but the pitfalls of even the most modern methods are revealed at a minimum in the still-limited success rates of wildcat oil and gas wells.

Both data distribution and inherent complexity affect confidence levels. Rarely do geologists have the luxury of a continuous outcrop or a closely-packed grid of boreholes. More commonly, stratigraphy depends on correlation of data from scattered outcrops, isolated boreholes, and sparse geophysical data. Even when data are relatively abundant, stratigraphic correlations are always interpretive—simply classifying strata chronometrically entails historiographic axioms. For creationists, the time emphasis in stratigraphy is itself troubling—the concept of a long, sufficiently uniform history which defines the goals and methods of stratigraphy is almost a polar opposite to the biblical record of catastrophic discontinuity. For example, Reed (2002) demonstrated that simply changing the interpretive paradigm from uniformitarian processes to the Genesis Flood created an entirely different history for the Palo Duro basin (Texas) apart from any substantive questioning of the details of existing stratigraphic correlations. For that reason alone, creationists must apply care in evaluating conclusions of uniformitarian scientists.

A still deeper and more disconcerting issue was raised at an epistemological level by Zeller (1964). In an effort to test the veracity of the stratigraphic methods used to correlate cyclic sediments, he created several random sedimentary columns which several geologists readily correlated with the real section he supplied. Not surprisingly, their stratigraphies differed, but they vigorously defended their respective stratigraphic columns.

> From the preceding story, it will be seen that our stratigraphic section, composed of randomly selected lithologies, does indeed show most of the characteristics that can be expected in a truly cyclic sequence. At this point the reader may wish to complain that the writer has gone too far in making up samples with which to taunt his colleagues. Let the reader be assured, however, that the writer's humble efforts at creating confusion are of truly minute proportions when compared to those of nature (pp. 635, 636).

In our many decades of applied stratigraphy, it has been our experience that this study and the few others like it are not widely taught in either academia or industry.

Zeller's results imply the impossibility of a unique stratigraphic solution. No correlation can be absolute, only more likely or less likely based on criteria such as predictive power. Some researchers have been perceptive enough to acknowledge this (Howe and Williams, 1994). Miall (1992) repeated Zeller's experiment using four randomly generated columns

[7] See Klevberg et al. (2003) for a more in-depth discussion of potential pitfalls, especially pp. 142–145.

and the Vail Curve. They all correlated extremely well. What many practitioners fail to realize is that only one fictitious correlation is required to discredit a deductive stratigraphic framework. This is the greatest weakness of imputing scientific certainty to historic reconstruction: the best reconstruction may be as far from the truth as the least tenable, and without the aid of an observer, we can never know. Viewed in that light, creationists operating within the biblical worldview appreciate the confidence available through its foundational truths, a confidence that those who depend on empiricism alone will never achieve.

Conclusions

Debate over the application of the geologic column within diluvial geology cannot escape the context of a larger debate over the nature of stratigraphy and its relationship to natural history. For nearly two centuries, stratigraphy has been held captive by the philosophical straitjacket of naturalism, which seeks to: (1) restrict reality to matter/energy; (2) restrict truth to science; and (3) subsume history within science. Creationists propose a bold program—the restoration of both science and history to their proper places within the biblical worldview. This will entail either the revision or rejection not simply of the geologic column, but of many of its associated axioms, methods, and conclusions of naturalism.

With regards to stratigraphy, creationists must first and foremost draw a clear and unambiguous distinction between science and history, insisting that natural history is at its root an historical pursuit, not a scientific one. In doing so, the myth of "prehistory" must be put to rest. Such a view does not reject the application of forensic methods to historical analysis; it merely places such efforts within a context that can guarantee the integrity of both science and history. Seeing natural history as a mixed question resolves the ongoing confusion characterized by such distinctions as "operations science" vs. "origins science." Instead, stratigraphy as a predominantly empirical enterprise is released from its burden of interpreting history and restored to its proper place as a forensic tool of natural history. As such, it does not bear the burden of historical uncertainty. Nor is it required to perform the tasks better relegated to revelation, theology, and philosophy—all inescapable ingredients in the mixed question approach.

Stratigraphy as an empirical discipline then becomes a point of commonality for advocates of different views about the geologic column within diluvial thinking, and even for advocates of either worldview. The methods of stratigraphy should then be evaluated by their descriptive potential, rather than by their support for the mythology of uniformitarianism. Uncertainty—the handicap of history, not the burden of science—could be more readily recognized, more easily tolerated, and dealt with openly rather than hidden behind authoritative dogmatism. If creationists pursue this kind of stratigraphy, they will realize their potential to profoundly influence thinking within the earth sciences as well as in the areas of origins and Earth's history that are presently the arena of the battle between Christianity and naturalism.

As the arguments unfold throughout the remainder of this book both for and against the geologic column, its applicability to diluvial geology, and its ultimate fate in the light of an ascendant creationist view of history; we urge the reader to keep sight of the wider implications for both stratigraphy and natural history within the biblical worldview.

References

CRSQ—*Creation Research Society Quarterly*

CENTJ—*Creation Ex Nihilo Technical Journal* (now *The Journal of Creation)*

Adler, M.J. 1965. *The conditions of philosophy*. Atheneum Press, New York.

———. 1993. *The four dimensions of philosophy*. Macmillan. New York.

Ancil, RE. 1983. The limits of human thought and the creation model. *CRSQ* 20:30–39.

———. 1985. On the importance of philosophy in the origins debate. *CRSQ* 22:114–123.

Bartlett, A.C. 1997. Sequence stratigraphy: value and controversy—for whom? *CRSQ* 34:5-22.

Bartz, P.A. 1987. If evolution is true—science is impossible! *Bible-Science Newsletter* 25(8):3.

———. 1989. When faith and science should differ. *Bible-Science Newsletter* 27(3):3.

———. 1992. Knowledge with humility. *Bible-Science News* 30(8):4,5.

Bates, R.L. and J.A. Jackson (editors). 1984. *Dictionary of geological terms.* Anchor Press, Garden City, NY.

Becker, C.L. 1932. *The heavenly city of the eighteenth-century philosophers.* Yale University Press, Clinton, MA.

Bergman, J. 1982. What is science? *CRSQ* 20:39–42.

Berthault, G. 1997. Sedimentation experiments: is extrapolation appropriate? A reply. *CENTJ* 11:65–70.

Brand, L.R. 1974. A philosophic rationale for a creation-flood model. *Origins* 1:73–83.

———. 1985. Can science and religion work together? *Origins* 12:71–88.

———. 1996. The paradigm of naturalism, compared with a viable alternative: a scientific philosophy for the study of origins. *Origins* 23:6–34.

Clark, G. 1978. *Religion, reason and revelation.* Craig Press, Nutley, NJ.

———. 1994. *Historiography secular and religious.* The Trinity Foundation, Jefferson, MD.

Clark, H.W. 1968. *Fossils, flood, and fire.* Outdoor Pictures, Escondido, CA.

Clark, J.D. 1976. Some philosophical implications of the theory of evolution. *Origins* 3:38–45.

Collingwood, R.G. 1956. *The idea of history.* Oxford University Press, London.

Copi, I.M. 1982. *Introduction to logic.* MacMillan Publishing Company, New York.

Davison, G.E. 1995. The importance of unconformity-bounded sequences in flood stratigraphy. *CENTJ* 9:223–243.

Frangos, A. 1991. The correct approach to scientific theories. *CRSQ* 28:17,18.

Froede, C.R., Jr. 1994. Sequence stratigraphy and creation geology. *CRSQ* 31:138–147.

———. 1995. A proposal for a creationist geological timescale. *CRSQ* 32:90–94.

———. 1997. The flood event/ice age stratigraphic boundary on the United States southeastern coastal plain. *CRSQ* 34:75–83.

———. 1998. *Field studies in catastrophic geology.* Creation Research Society Books, St. Joseph, MO.

Garner, P. 1996. Where is the Flood/post-Flood boundary? Implications of dinosaur nests in the Mesozoic. *CENTJ* 10(1):101–106.

Garton, M. 1996. The pattern of fossil tracks in the geological record. *CENTJ* 10:82–100.

Gill, G.H. 1996. A sufficient reason for false Rb-Sr isochrons. *CRSQ* 33:105–108.

Gould, S.J. 1997. *Time's arrow time's cycle: myth and metaphor in the discovery of deep geological time.* Harvard University Press, Cambridge, MA.

Helmick, L.S. and D.P. Baumann. 1989. A demonstration of the mixing model to account for Rb-Sr isochrons. *CRSQ* 26:20–23.

Holt, R.D. 1996. Evidence for a late Cainozoic flood/post-flood boundary. *CENTJ* 10:128–167.

Hoffman, E.J. 1993. Evolutionism: an oxymoron. *CRSQ* 30:12–16.

Howe, G.F. and E.L. Williams. 1994. The evolution of geological origins theories: part I—the Haymond interbeds, Marathon Basin, Texas. *CRSQ* 31:25-31.

Hull, G.A. 1989a. Testing our truth claims —Part I: rational coherence. *Bible-Science Newsletter* 27(4):1,2.

———. 1989b. Testing for truth – Part II: empirical adequacy. *Bible-Science Newsletter* 27(5):1,5.

Jeletzky, J.A. 1978. *Causes of cretaceous oscillations of sea level in western and arctic Canada and some general geotectonic implications.* Geological Survey of Canada Paper 77–18.

Johansson, Åke. 1993. Interview with Åke Johannson. *Genesis* 2-93:23.

Johnson, P.E. 1991. *Darwin on trial.* Regnery Gateway, Washington, D.C.

———. 1993. Creator or blind watchmaker? *First Things*, January, 1993:8–14. Downers Grove, IL.

———. 1995. *Reason in the balance.* InterVarsity Press, Downers Grove, IL.

Johnson, R. 1993. Hur gammal är jorden? Översikt över olika dateringsmetoder (How old is the world? Overview of various dating methods). *Genesis* 2-93:18–22.

Klevberg, P. 1999. The philosophy of sequence stratigraphy Part I–philosophic background. *CRSQ* 36(2):72–80.

———. 2000a. The philosophy of sequence stratigraphy Part II – application to stratigraphy. *CRSQ* 37(1):36–46.

———. 2000b. The philosophy of sequence stratigraphy Part III – application to sequence stratigraphy. *CRSQ* 37(2):94–104.

Klevberg, P., M.J. Oard, and R. Bandy. 2003. Are paleosols really ancient soils? *CRSQ* 40(3):134–149.

Kofahl, R.E. 1986. Correctly redefining distorted science: a most essential task. *CRSQ* 23:112–114.

———. 1989. The hierarchy of conceptual levels for scientific thought and research. *CRSQ* 26:12–14.

Lammerts, W.E. 1984a. Recorded instances of wrong-order formations: a bibliography—Part I. *CRSQ* 21:88.

———. 1984b. Recorded instances of wrong-order formations: a bibliography—Part II. *CRSQ* 21:150.

———. 1985a. Recorded instances of wrong-order formations: a bibliography—Part III. *CRSQ* 21:200.

———. 1985b. Recorded instances of wrong-order formations: a bibliography—Part IV. *CRSQ* 22:127.

———. 1986a. Recorded instances of wrong-order formations: a bibliography—Part V. *CRSQ* 22:188,189.

———. 1986b. Recorded instances of wrong-order formations: a bibliography—Part VI. *CRSQ* 23:38.

———. 1986c. Recorded instances of wrong-order formations: a bibliography—Part VII. *CRSQ* 23:133.

———. 1987. Recorded instances of wrong-order formations: a bibliography—Part VIII. *CRSQ* 24:46.

Lemon, R.R. 1991. *Principles of stratigraphy*. Merrill Publishing Company, Columbus, OH.

Lillo, C.L. 1987. Enlightenment or endarkenment? *CRSQ* 24:19–23.

Lumsden, R.D. 1992. Error and worse in the scientific literature. *CRSQ* 29:127–132.

Malcolm, D. 1997. A philosophical attempt to define science. *CENTJ* 11:167–180.

McGhee, L.A. 1987. The metaphysics of modern science. *CRSQ* 24:138–141.

Mehlert, A.W. 1986. Diluviology and uniformitarian geology—a review. *CRSQ* 23:104–109.

———. 1993a. An evaluation of the John Woodmorappe flood geology model—part I. *CRSQ* 30:77-86.

———. 1993b. An evaluation of the John Woodmorappe flood geology model—part II. *CRSQ* 30:149–159.

Miall, A.D. 1986. Eustatic sea level changes interpreted from seismic stratigraphy: a critique of the methodology with particular reference to the North Sea Jurassic record. *American Association of Petroleum Geologists Bulletin* 70:131–137.

———. 1992. Exxon global cycle chart: an event for every occasion? *Geology* 20:787–790.

Middelmann, U. and A.E. Wilder-Smith. 1980. Den som tenker må tro (He who thinks must believe). Prokla-Media, Ottestad, Norway.

Molén, M. 1991. Vårt ursprung? (Our origin?) Umea FoU, Umea, Sweden. Cited in R. Johnson.

Moreland, J.P. 1989. *Christianity and the nature of science*. Baker Book House, Grand Rapids, MI.

Morris, H.M. 1996. The geologic column and the flood of Genesis. *CRSQ* 33:49–57.

Mortenson, T. 2004. *The great turning point: the church's catastrophic mistake on geology before Darwin*. Master Books, Green Forest, AR.

North American Commission on Stratigraphic Nomenclature. 1983. North American stratigraphic code. *American Association of Petroleum Geologists Bulletin* 67(5):841–875.

Parkinson, N. and C. Summerhayes. 1985. Synchronous global sequence boundaries. *American Association of Petroleum Geologists Bulletin* 69:685–687.

Pearcey, N.R. 1987. A science of origins: an interview with Charles Thaxton. *Bible-Science Newsletter* 25(1):6–9.

Pearcey, N.R. and Johnson, P.E. 2004. Total truth: liberating Christianity from its cultural captivity. Crossway Books, Wheaton, IL.

Plantinga, A. 1990. *The twin pillars of Christian scholarship*. Calvin College and Seminary. Grand Rapids, MI.

———. 1997a. Methodological naturalism? *Origins & Design* 18(1):18–27.

———. 1997b. Methodological naturalism: Part 2. *Origins & Design* 18(2):22–34.

Reed, J.K. 1998 Demythologizing uniformitarian history. *CRSQ* 35(3):157–165.

———. 2000a. Historiography and Natural History. *CRSQ* 37(3):160–175.

___ (editor). 2000b. *Plate tectonics: a different view.* CRS Books, Chino Valley, AZ.

———. 2001. *Natural history in the Christian worldview: foundation and framework*. Creation Research Society Books, St. Joseph, MO.

———. 2002. Reinventing stratigraphy at the Palo Duro Basin. *CRSQ* 39:25–39.

Renard, K. 1993. Den fria viljan (Free will). *Genesis* 1-

93:20–25.

Riemen, W.P. 1987. The non-material hypothesis and its implications for modern science. *CRSQ* 23:141–145.

Ritland, R. 1981. Historical development of the current understanding of the geologic column: part I. *Origins* 8:59–76.

———. 1982. Historical development of the current understanding of the geologic column: part II. *Origins* 9:28–50.

Robinson, S.J. 1995. From the flood to the exodus: Egypt's earliest settlers. *CENTJ* 9:45–68.

———. 1996. Can flood geology explain the fossil record? *CENTJ* 10:32–69.

———. 1997. The geological column: a concept fundamental to flood geology. *Origins* 23:14–30.

Schaeffer, F.A. 1968a. *Escape from reason*. In, *The complete works of Francis A. Schaeffer—volume I: a Christian view of philosophy and culture*. Crossway Books, Westchester, IL.

———. 1968b. *The God Who is there*. In, *The complete works of Francis A. Schaeffer—volume I: a Christian view of philosophy and culture*. Crossway Books, Westchester, IL.

———. 1972. *He is there and He is not silent*. In, *The complete works of Francis A. Schaeffer—volume I: a Christian view of philosophy and culture*. Crossway Books, Westchester, IL.

———. 1976. *How should we then live?* In, *The complete works of Francis A. Schaeffer—volume V: a Christian view of the West*. Crossway Books, Westchester, IL.

Schlossberg, H. 1983. *Idols for destruction*. Thomas Nelson Publishers, Nashville, TN.

Schoepflin, G.L. 1982. Perceptions of the nature of science and Christian strategies for a science of nature. *Origins* 9:10–27.

Sire, J.W. 1988. *The universe next door.* InterVarsity Press, Downers Grove, IL.

Sloss, L.L. 1963. Sequences in the cratonic interior of North America. *Geological Society of America Bulletin* 74:93–114.

Snelling, A.A. 1995. The failure of U-Th-Pb 'dating' at Koongarra, Australia. *Creation Ex Nihilo* 9:71–92.

———. 1992. The case of the 'missing' geologic time. *Creation Ex Nihilo* 14(3):30–35.

———. 1998. The cause of anomalous potassium-argon "ages" for recent andesite flows at Mt. Ngauruhoe, New Zealand, and the implications for potassium-argon "dating." In Walsh, R.E. (editor). *Proceedings of the Fourth International Conference on Creationism*. pp. 503–525. Creation Science Fellowship, Pittsburgh, PA.

Snelling, A.A., M. Ernst, E. Scheven, J. Scheven, S.A. Austin, K.P. Wise, P. Garner, M. Garton, and D. Tyler. 1996. The geological record. *CENTJ* 10:333,334.

Sproul, R.C., J. Gerstner, and A. Lindsley. 1984. *Classical apologetics*. Academie Books, Grand Rapids, MI.

Stark, R. 2003. *For the glory of God*. Princeton University Press, Princeton, NJ.

Sunderland, L.D. 1986. The geologic column: its basis and who constructed it. *Bible-Science Newsletter* 24(12):1,2,5,6,14.

Taylor, I.T. 1991. *In the minds of men: Darwin and the new world order*. TFE Publishing, Toronto, ON.

Thorne, J. and A.B. Watts. 1984. Seismic reflectors and unconformities at passive continental margins. *Nature* 311:365–368.

Titus, H.H., M.S. Smith, and R.T. Nolan. 1979. *Living issues in philosophy.* Wadsworth Publishing Company, Belmont, CA.

Vail, P.R., R.M. Mitchum, Jr., and S. Thompson, III. 1977. Seismic stratigraphy and global changes of sea level, Part 4: global cycles of relative changes of sea level. In Payton, C.E. (editor). *Seismic stratigraphy—applications to hydrocarbon exploration*. pp. 83–97. American Association of Petroleum Geologists, Tulsa, OK.

Vail, P.R., F. Audemard, S.A. Bowman, P.N. Eisner, and C. Perez-Cruz. 1991. The stratigraphic signatures of tectonics, eustasy and sedimentology—an overview. In Einsele, G., W. Ricken, and A. Seilacher (editors). *Cycles and events in stratigraphy.* pp. 617–659. Springer-Verlag, New York.

Van Wagoner, J.C., H.W. Posamentier, R.M. Mitchum, Jr., P.R. Vail, J.F. Sarg, T.S. Loutit, and J. Hardenbol. 1988. An overview of the fundamentals of sequence stratigraphy and key definitions. In Wilgus, C.K., B.S. Hastings, H. Posamentier, J. Van Wagoner, C.A. Ross and C.G. St. C. Kendall (editors). *Sea-level changes: an integrated approach.* pp. 39–45. Society of Economic Paleontologists and Mineralogists Special Publication No. 42, Tulsa, OK.

Vardiman, L, A.A. Snelling, E.F. Chaffin. 2000. *Radioisotopes and the age of the earth.* Institute for Creation Research, El Cajon, CA and Creation Research Society, St. Joseph, MO.

Verbrugge, M. 1985. The legacy of Duyvene De Wit for creationist biology: Part III—the cosmonomic philosophy: a Christian alternative to evolutionism. *CRSQ* 21:182–184.

Walker, R.G. 1990. Facies modeling and sequence stratigraphy. *Journal of Sedimentary Petrology* 60(5):777–786.

Walker, T. 1996. The Great Artesian Basin, Australia. *CENTJ* 10:379–390.

Whitcomb, J.C. and H.M. Morris. 1961. *The Genesis flood*. Presbyterian and Reformed Publishing Company,

Philadelphia, PA.

Williams, E.L. 1992. Cavern and speleothem formation—science and philosophy. *CRSQ* 29:83–84.

Winchester, S. 2001. The map that changed the world. HarperCollins Publishers, Inc., New York.

Woodmorappe, J. 1978. A diluvian interpretation of ancient cyclic sedimentation. *CRSQ* 14(4):189–208.

———. 1980. An anthology of matters significant to creationism and diluviology: report 1. *CRSQ* 16(4):209–219.

———. 1981. The essential nonexistence of the evolutionary-uniformitarian geologic column: a quantitative assessment. *CRSQ* 18:46–71.

———. 1983. A Diluviological treatise on the stratigraphic separation of fossils. *CRSQ* 20:133–185.

———. 1996. Studies in Flood geology: clarifications related to the "reality" of the geological column. *CENTJ* 10(2):279–290.

———. 1999a. The geologic column: does it exist? *CENTJ* 13(2):77–82.

———. 1999b. *The mythology of modern dating methods.* Institute for Creation Research, El Cajon, CA.

Zeller, E.J. 1964. Cycles and psychology. *Kansas Geological Survey Bulletin* 169:631–636.

Øhrstrøm, P. 1990. Den frie vilje og videnskaben (Free will and science). *Origo* 7(4):14–19.

Authors

John K. Reed, Ph.D., has served in the Creation Research Society as a board member and the geology editor for the Creation Research Society Quarterly. He is the author of five books and numerous articles.

Peter Klevberg, B.S., is a geologist who has written articles about stratigraphy, paleosols, and diluvial deposits in the northern Rocky Mountains.

Carl. R. Froede, Jr., B.S., P.G., is a lifetime member of the Creation Research Society. He is the author of one book and numerous geological articles, many within the Creation/Flood framework.

The authors would like to acknowledge the help in preparing this chapter from a series of papers about sequence stratigraphy by Peter Klevberg (1999; 2000a; 2000b) published in the *CRSQ*.

Appendix: Sequence stratigraphy

Sequence stratigraphy, a recent major innovation, emphasizes sedimentary sequences. Sequences are defined by their bounding surfaces (usually unconformities) rather than more traditional bedding or faunal markers. Although sequences were recognized by Sloss (1963) in the North American midcontinent, the concept may date back to Walther (Berthault, 1997, p. 67). However, the method arose from 20th-century advances in the acquisition, processing, and display of seismic reflectors. These are grouped into distinct packages bounded by unconformities, and then interpreted as transgressive-regressive cycles, usually attributed to eustasy. These sea level changes are thought to provide a global "clock" for correlation, and the Vail Curve (Vail et al., 1977) or similar eustatic curves are commonly added to presentations of the geologic column. There are variations on the theme. According to Walker (1990, p. 780):

> There are currently at least four stratigraphies that attempt to subdivide rocks into genetic packages based on bounding unconformities or discontinuities. They are largely conceptual.... They all derive from seismic stratigraphy...

Bartlett (1997) lists these as: (1) classical sequence stratigraphy (the Exxon approach), (2) genetic stratigraphic sequences (the Galloway approach), (3) allostratigraphy,[8] and (4) transgressive-regressive cycles.

Sequence stratigraphy's empirical potential has recommended it to some creationists.

> It is suggested that this concept be adopted for interpreting Flood geology for the following reasons: (1) [unconformity-bounded sequences] are defined by physical boundaries (unconformities), that is, they are lithostratigraphic (as opposed to time stratigraphic [evolutionary]); (2) each sequence represents the spatial and temporal distribution of a complete depositional 'cycle,' that is, a geologic history; and (3) each sequence refers to a cluster of strata of varying but continuous (relative) age, that is, relatively continuous deposition (within the sequence/depocenter) [Davison (1995, p.228].

Some creationists believe it can be used absent any historic bias:

> Sequence stratigraphy represents a breakout from the intellectual-philosophic Bastille of uniformitarian time (Bartlett, 1997, p. 12).

However, the interpretation of seismic sequences as synchronous eustatic changes relies (at a minimum) on evolutionary chronology in the column.

> *...the geologist must define time-equivalent rock packages that are genetic intervals* (Vail et al., 1991, p. 617).

> Sequence stratigraphy is the study of rock relationships within a *chronostratigraphic framework of repetitive, genetically related strata...* (Van Wagoner et al. 1988, p. 39, emphasis added).

Note how seamlessly evolutionary fossil succession is woven into sequence stratigraphy. Vail et al. (1991) proposed six steps; at least four of which require acceptance of radiometric dating and evolutionary biostratigraphy. The global eustatic curve, based on uniformitarian chronology, is implicit in sequence stratigraphy (Froede, 1994). Thus, none of the sequence stratigraphy approaches, excepting allostratigraphy, is purely descriptive. Allostratigraphy, unfortunately, is applicable only to particular sedimentary basins where discontinuities can be

[8] Only allostratigraphy is recognized by the North American Commission on Stratigraphic Nomenclature, and some authors do not include allostratigraphy with sequence stratigraphy. Allostratigraphy does differ from the other methods in its descriptive nature (it is based solely on physical discontinuities between units), but its obvious similarities justify its inclusion with other sequence stratigraphic approaches.

adequately determined to map individual units. In general, sequence stratigraphic methods are inherently subjective or nonempirical in their dependence upon genetic assumptions about history.

Other creationists have expressed reservations.

> Recently, claims have been made of the possibility of highly detailed global chronostratigraphic correlations based upon sequence stratigraphy. Miall (1992) provides a devastating critique of these claims. He shows that the range of errors in biostratigraphy, magnetostratigraphy and isotopic dating are such that the precision claimed is a physical impossibility. Furthermore, he highlights the role of circular reasoning and fortuitous correlations in, for instance, the belief that there are 40 discernible global events within just the Cretaceous Period.... There is also at least some element of circular reasoning in even the first-order cycles of the Vail curve. The times inferred for even first-order global regressive sequences contradict each other, and depend upon which database is used. Of course, the sea-level curves should not be directly imported into Flood geology. For instance, 'marine regressive' sequences need not be accepted as such in Flood geology, but may be interpreted as giant reverse-graded beds laid down by Flood waters that are increasing in velocity at that point of deposition (Woodmorappe, 1996, p. 280).

Even uniformitarians have voiced doubts. Parkinson and Summerhayes (1985, p. 686) expressed an early concern that has since become more widespread:

> The danger in using the global curve to explain stratigraphic breaks is that it impairs thinking about local controls. We think it is time to return to the unfettered examination of individual basins on their own merits.

Roger Walker (1990, p. 777, 780) noted:

> The significance accorded the contacts is one of the main problems in stratigraphy and sedimentology... There are currently at least four stratigraphies that attempt to subdivide rocks into *genetic packages* based on bounding unconformities or discontinuities. *They are largely conceptual*, with little or no consideration of 1) scale of application, 2) actual geological examples... or 3) the relationship between the different schemes (emphasis added).

Froede (1994) pointed out other weaknesses in the Vail curve, including implicit confidence in the Milankovitch mechanism and Walther's Law. Various limitations, including the issue of tectonic overprinting, remain. The direct connection between sequence stratigraphy as practiced and various fallacious evolutionist methods is inescapable. Klevberg (2000b) identified the following problems with sequence stratigraphy:

- Subjectivity in identification of unconformities and correlative conformities.
- Subjectivity in reconstruction of paleoenvironments and systems tracts.
- Subjectivity in identification of many sequence boundaries.
- Assumption of relatively constant sediment input.
- Assumption of eustasy and Milankovitch mechanism.
- Assumption of subaerial erosion surfaces incompatible with Deluge.

Thus, while some of the methods employed by sequence stratigraphers may prove useful, it is not free of the influence of the usual suspects.

Glossary

Allostratigraphy—a stratigraphic scheme which categorizes rock units base on their geometric (spatial) relations within a sedimentary basin

axiom—an assumed or accepted principle, often regarded as self-evident

biostratigraphy—correlation of rock units based on fossil content

chronostratigraphy—a scheme of time units devised for interpretation of geologic (stratigraphic) data

epistemology—discipline concerned with the theory of knowledge or how one can know

eustasy—global sea level and its fluctuations in time

genetic—pertaining to mode of origin

lithostratigraphy—correlation of rock units based on lithology (rock type)

lithodemic stratigraphy—stratigraphy of nonstratiform rocks based on relative dating —criteria (i.e. cross-cutting relations)

metaphysics—study of the nature of ultimate reality

neocatastrophism—modern emphasis on the importance of catastrophic geologic events

positivism—the belief that only empirical propositions are meaningful and that metaphysics is impossible

stratigraphic sequence—a package of conformable strata, often cyclic; in sequence stratigraphy, the term is applied at a regional scale, usually comparable to a group; otherwise, it is usually a smaller sedimentary package of formation or lesser rank

unconformity—a contact between rock units that does not conform to the geometry or fabric of the adjacent beds; it often represents an erosion surface

uniformitarianism—the doctrine that geologic processes in the past did not differ substantially from those observed in the present

seismic data—subsurface information obtained in the form of wave transmission rates and patterns of reflection and refraction; other geophysical methods are often used in conjunction with seismic

Vail Curve—the curve of relative sea level or eustasy

lithofacies—a mappable rock body distinguished by lithology

paleoclimatology—study (or speculation) of ancient climates, inferred from various data (lithofacies, paleogeography, fossils, etc.)

paleoecology—study/speculation relative to ecology of an area in ancient times

paleoenvironments—past environments, generally inferred from lithofacies, fossils and other data

paleogeography—study of geography at points in the past, often inferred from facies changes

paleosol—a "fossil soil" or lithified soil horizon

Walther's Law—doctrine that a vertical facies sequence represents a series of laterally adjacent depositional environments in the same associations as are observed today

four

Accept the Column, Reject the Chronology

David J. Tyler

Harold G. Coffin

Abstract

The status of the geological column has led to strong differences of opinion among diluvialists. This chapter seeks to explain why the authors consider the geological column concept to be foundational for developing an understanding of the geological history of Planet Earth. The concept is of great assistance in structuring thinking about field observations of rock strata and their associated fossils.

Historical background

Local stratigraphic order of strata is universally recognized and generally does not create controversy. Nicholaus Steno, the Father of Geology, is famous for the sections he drew of complex rock layers in his local area of Tuscany, and for originating the concept of superposition of strata (Cutler, 2003). He used his observations as a basis for reconstructing a history of how the strata came to be fractured. In his mind, the Genesis Flood laid down distinct layers of different sediment types, which later hardened and experienced deformation. The concept of superposition linked to a geological history that has stood the test of time and has proved to be a necessary ingredient in the tool-kit of geologists.

Many pioneers built on this foundation, but the next major breakthrough for stratigraphy came with the work of William Smith, often referred to as the "Father of British Geology." In his working life, Smith was a professional who got involved with canal building, surveying mines, advising mine owners and so on. He did not have independent means that allowed him to pursue geology as an intellectual interest. He needed to find tools that were useful in his work. He found patterns in the rocks he was excavating: the enclosed fossils often provided a signature which distinguished a particular rock stratum. This proved to be useful as a mapping tool, and it launched William Smith on what became his greatest achievement: mapping the geology of England and Wales.

Numerous people recognized that William Smith was doing something of great significance, but his working class background was a factor in the delayed publication of his maps. He did meet other geological leaders, but was never received into the community of gentlemen scholars. It became apparent that Cuvier and Brogniart were doing something similar in the Paris Basin, publishing their work in 1812, before Smith's map appeared in 1815.

During the development of Smith's "law of faunal succession," the leading geological minds were occupied with rather different issues. In Germany, Werner had proposed his model of precipitation from a primeval ocean to produce stratified rocks. The English-speaking world came to adopt the stratigraphical terms: Primary, Transitional, Secondary and Tertiary. In the time when Smith was developing his ideas, there was controversy, but not about the broad brush stratigraphy. Rather, it was about whether the rocks were precipitated from water (as Werner maintained) or whether some of these rocks had solidified from a molten state (as advocated by Hutton). Linked to this were questions about timescales: Werner's approach allowed, but did not require, short timescales whereas Hutton's required long geological ages.

Hutton made the crucial break with time's arrow: he proposed that "there is no vestige of a beginning and no prospect of an end." The relentless cycling of the 'Earth machine' dominated Hutton's geology. The machine was brought into being by a Deistic god, set in motion and automatically sustained by the operation of observable geologic processes to fulfill the divine purpose. As long as the current order of natural law was maintained, the geologic cycle would continue. Though Hutton is thought to have derived his ideas after much pondering over field evidence, in reality, he first developed a theoretical model, which he then sought to relate to field evidences (Gould, 1987). By developing such ideas, Hutton was the first to take the concept of "deep time" beyond the realm of speculation and, consequently, is widely regarded as the Father of modern geology.

While Hutton (in Scotland) and Werner (in Germany) championed different schools of thought, a community of English stratigraphers was emerging. These people were the real empiricists. They were interested in documenting the rocks. Recognizing different lithologies within the Primary, Secondary and Tertiary divisions, and using the laws of superposition and faunal succession, these field geologists put together a lithological column representing the rocks in England and what they knew of Europe. Rupke (1983, p. 125) wrote:

> The 1821 table of international correlation and classification was remarkably advanced. The only major deficiencies were the lack of a subdivision of the Transitional Series and of a sharp Permo-Triassic boundary; apart from this all major formations or systems of the definitive version of approximately 1840 were recognized:
>
> (1) Transition limestone and Greywacke,
> (2) Old Red Sandstone,
> (3) Carboniferous Limestone,
> (4) Coal Measures,

(5) New Red Conglomerate,
(6) New Red Sandstone and Magnesium Limestone,
(7) Oolite or Jura Limestone,
(8) Chalk and Greensand,
(9) Tertiary,
(10) Diluvium,
(11) Alluvium.

Subsequent European exploration brought greater clarity, with confirmation of the lithological sequences supplemented by some additional named horizons. During this period, the influence of Charles Lyell was such that catastrophism declined and the uniformitarianism of Hutton and Lyell became dominant. The results of field geology were consistently presented within the context of long geological ages.

With the publication in 1830–1833 of Charles Lyell's *Principles of Geology*, a major change of mindset occurred within the geological community. Lyell continued the tradition of Hutton but with a strong philosophical commitment to uniformity in all its possible meanings: the uniformity of natural law, the uniformity of process, the uniformity of rate and the uniformity of state (Gould, 1987). Unfortunately, Lyell did not draw clear distinctions between these different meanings, and yet all were defined to be essential to the development of geology as a science. As an example, although Hutton was prepared to accept catastrophic upheavals of the Earth's crust during phases of mountain building, Lyell argued against this. His commitment to uniformity of process, rate and state dominated his thinking. Lyell's identification of science with all meanings of uniformity was extraordinarily influential and "uniformitarianism" has been an unhealthy constraining bias in the development of geological thought since then.

In addition, Lyell is remembered for the division of the Tertiary in 1833 in accordance with the proportions of living/extinct fossils: 3% in Eocene, 17% in Miocene, 50–67% in Pliocene. This was a significant deviation from the earlier empirical field geology, but was considered justified because the Tertiary strata is much more fragmented and localized than the earlier systems. Whereas earlier work built on direct observations of stratigraphical relationships, Lyell's division of the Tertiary used a theoretical tool to interpret and bring order to the limited field data.

Nevertheless, the strong tradition established by the English stratigraphers was maintained. Hugh Miller, when he was elected in 1852 as President of the Royal Physical Society of Edinburgh, wrote in his inaugural address:

> All over the globe, too, in the great Palaeozoic division, the Carboniferous system is found to overly the system of the Old Red Sandstone, and that, in turn, the widely developed Silurian system. It is not less certain, that in the Secondary division, the Triassic deposits are overlain by the Oolitic ones, and both by the Cretaceous; nor yet, that in the Tertiary division, the beds of the Pliocene, with their large percentages of existing shells, as exemplified in the Red and Coraline Crags, belong to a greatly later period than that old Eocene age represented by the extinct shells and strange mammals of the Paris basin and the London Clay. There is no human history more definitely ranged into centuries than the geological into periods and epochs; nor is the certainty less great, or the chance of transposition in any degree less slight, in the one case than in the other. For, respecting at least the main geologic systems, their order of succession, and the organisms which they contain, the evidence is as positive and conclusive as it is regarding any piece of human history whatever. (Miller, 1852, p. 276).

In this same address, Miller pointed to the importance of the law of faunal succession in deciphering stratigraphy. As a Christian and one of the leaders of the Free Church of Scotland, he explained his own position on origins. He accounted for the faunal successions by postulating successive creations of God through the geological ages.

> When it was first ascertained by the father of English geology, William Smith—a man not yet more than twelve years dead—that the Oolitic beds of England have always a uniform order of succession, and that this uniformity is attended by a certain equally uniform succession of groups of fossils, could it be once inferred that he was laying hold of a principle which, in the course of a single age, was destined marvellously to unlock the past history of our planet, and to acquaint us with God's doings upon it, as the Creator of all, for myriads of ages ere He had first breathed the spirit of life into human nostrils, or man had become a living soul? (Miller, 1852, p. 275).

Rationale for establishing correlations between strata

In our historical review, it is clear that local geologic successions can be recognized as a matter of direct observation. Individual outcrops can demonstrate superposition relationships. For example, at Cayton Bay (Figure 1), on the Yorkshire coast (England), there is no difficulty recognizing that a calcareous sandstone (Red Cliff Rock Member) is overlain by a mudstone (Oxford Clay Formation), which in turn is overlain by another calcareous sandstone (Lower Calcareous Grit Formation).

Observations made from vertical sections like this are greatly extended by walking along strike, as individual beds can be traced over land to find the stratigraphical relationships with beds above or below. Thus, the "Jurassic Coast" World Heritage

Figure 1. Cayton Bay section. A local column exposed at the Yorkshire coast (England). See text for further details.

Figure 2. Jurassic Coast monument. The Geoneedle is at the far west of the World Heritage Coast section (England), with representative samples of rocks in the section built into the monument.

locality along the coast of Devon and Dorset (England) allows the sequence from Triassic redbeds to Cretaceous chalk overlain by Tertiary beds to be examined. A monument (Figure 2) near Sidmouth (Dorset) has samples of many members of this succession. Anyone doubting the reality of this local section can check it out by walking along the coast and noting the outcrop and superposition relationships of the different rock units.

Extensive correlations between local sections can be built up by careful observations of field relations, lithologies and fossil content. This was done for England and Wales by the year 1821 (as indicated above) and it was apparent in William Smith's map of 1815. Thus, the geology of England and Wales can be represented by a single local column.

Is it possible to extend the correlation beyond islands, beyond regions and even beyond continents? This is the crucial issue addressed in this chapter, and the authors set out a case for answering in the affirmative.

Connecting local columns

One of the features of the rock record is that certain lithologies, with characteristic fossils, are remarkably persistent. Thus, rock units in England are very similar to rocks in Germany and France, but direct evidence of continuity is difficult because they are separated by the English Channel. Nevertheless, similar successions and their distinctive fossils are found and geologists concluded that there must be a physical connection. Direct confirmation of this approach has come recently, when boreholes to prove the geology of the rocks in the English Channel were drilled to assist with the construction of the Channel Tunnel. Lateral persistence was amply confirmed.

By 1840, the amended geological column of England and Wales was demonstrably a geological column for Europe. Geologists were aware that the column was certainly not complete in any specific area, but each area provided a sampling of the whole, generally maintaining the same order.

Later in the 19th century, out-of-sequence strata were recognized, but these were usually in mountainous terrains and always associated with folding and faulting of rocks. While normal and reverse faulting frequently brought different rock types into contact, it was the low angle thrust faults that created the strongest contrasts and significant examples of out-of-sequence strata. In such cases, mapping work to assess the lateral displacement was usually sufficient to identify their source.

In the United States of America, exploration demonstrated that although many of the stratigraphical units were even more extensive than Europe, recognizably similar successions could be found. Thus, the Colorado Plateau, which incorporates parts of the states of Arizona, Colorado, and Utah, is one of the best locations in the world to study stratigraphy. This high plateau is cut by many deep canyons that expose the strata (Figure 3). The Grand Canyon in Northern Arizona opens up a window into the Precambrian and the Paleozoic (only the Silurian and Ordovician periods are missing). A few km to the north, the Vermillion Cliffs signal the beginning of Mesozoic strata that overlie the Paleozoic. Another step (also a few km further north) adds the upper beds of the Mesozoic and the Lower Cenozoic. This stratigraphy can be easily observed as a series of "local sequences" that can be connected together by walking along strike and the overall succession has been verified by examination of numerous oil well cores.

Figure 3. Basal conglomerate located between the Precambrian and the Cambrian in the Grand Canyon of Colorado River, Arizona.

Using the same principles, the terrain northeast of Yellowstone National Park also exposes strata from the Precambrian to the Lower Tertiary (with glacial Pleistocene at the very top) although the sequence is not quite as easy to observe as in the Grand Canyon region.

Not all strata can be correlated by fossil content because some strata have no fossils or those present are not guide or index fossils that distinguish the bed from other beds. Nevertheless fossil content often is a unique and useful tool for the correct stratigraphic placement of a bed. Two examples from North America are the Triassic Shinarump Conglomerate (Figure 4) with petrified wood and trees and the Cretaceous Mowry shale with abundant fossil fish scales (Figure 5). The former is a relatively thin bed of coarse terrestrial deposits that covers parts of several mid-western states. The latter covers several northern, mid-western, Rocky Mountain states. Within this bed of black shale, a band two or three meters thick contains countless numbers of fossilized fish scales. Under normal conditions the scales of dead fish do not last long. The presence of millions of well-preserved fossil fish scales in the Mowry Shale must represent a situation involving rapid wide-spread deposition.

This story has been repeated many times. Local sections consistently confirm the Geological column concept. Near the base of the fossiliferous rock sequences (Paleozoic rocks), the dominant faunas are marine, with trilobites, nautiloids, graptolites and certain forms of coral. Towards the upper Paleozoic are major coal-bearing strata. Above these are red-beds largely devoid of macrofossils, followed by a variety of terrestrial and marine strata, often bearing witness to the presence of dinosaurs (Figure 6). This culminates in chalk, a very extensive and distinctive rock type, with its own suite of characteristic

Figure 4. Shinarump Conglomerate, a wide-spread formation exposed in several mid-western States.

fossils. Above this are Tertiary beds, typically less persistent, with fossil mammals and birds (but no dinosaurs).

This general sequence comes close to being a global phenomenon. Rarely can direct correlations be made across the ocean floors, but the patterns on either side are remarkably similar. One of the affirmations made by Whitcomb and Morris (1961) about the geological column was that the column concept did not fit the rest of the world, but they gave no supporting evidence for this apart from out-of-sequence fossils and a quotation from a petroleum geologist (p. 209). In our experience, these 'out-of-sequence' rocks are not laterally extensive, and are typically associated with mountain belts where there are numer-

Figure 5. A slab of Mowry Shale showing several fossil fish scales.

Figure 6. The Morrison Formation in which dinosaur remains are often found.

Figure 7. Heart Mountain in Wyoming with a block of the Mississippian Madison Formation sitting on the Tertiary Willwood Formation.

ous evidences of tectonic activity (folding, faulting, brecciation and overthrusting).

A classic and unusual example of "out-of-order" stratigraphy is seen in the Heart Mountain Thrust Fault. The top of Heart Mountain (Figure 7) in the Big Horn Basin in the state of Wyoming is capped by Mississippian Madison sediments. Based on the stratigraphy visible in the Rocky Mountains to the west, the Madison Formation should be sitting on Lower Paleozoic. This is not the case however, for Heart Mountain. The Mississippian cap sits on Tertiary sediments. Coal and oil are present in the Big Horn Basin. Oil wells have been drilled in the area around Heart Mountain. These wells have passed through Mesozoic and Paleozoic strata (in correct order) *including Mississippian Madison! Thus Madison beds are present both on top of the mountain and also several thousand feet below.*

Is this a case against the reality of the geological column as argued by some creationists? Or is there some other explanation? The Heart Mountain Thrust Fault has been studied extensively. Evidences clearly point to the movement of a large block of Paleozoic Madison eastward into the Big Horn Basin. Movement of this large segment of Paleozoic Madison caused it to break into smaller blocks, one of which became the top of Heart Mountain. Because the strata of the mountains were at higher elevation than those in the Big Horn Basin, some of the blocks moved out onto "younger" sediments. Beutner and Gerbi (2005, p. 724) have interesting comments on the mechanism:

> High CO_2 pressure drastically reduced friction along the fault and allowed continuation of catastrophic movement, probably initiated by a volcanic or phreatomagmatic explosion, resulting in very large displacement on a low-dipping surface. Earlier slower sliding may have occurred but final emplacement was rapid (minutes) and spectacular.

Thrust faults are present in most major mountain ranges around the world. The explanation for "out-of-order" strata usually is explainable by movements of rock due to mountain building activities. That is the case for the "out-of-order" block on the top of Heart Mountain. Detailed mapping of thrust blocks and close examination of thrust planes means that 'bedding plane' interpretations are not tenable. This should be regarded as "good news" for diluvialists. The catastrophist nature of overthrusting has long been one of the missing ingredients in the diluvialist challenge to uniformitarianism.

Having noted these things, we are not implying that the different continental masses are geological clones of each other. In the personal experience of one of us (HGC), the stratigraphy of South America, especially in Brazil and Peru, appears to be somewhat different from that in North America. The column there appears to be much abbreviated. We recognize that in some areas, the relation of the stratigraphy to the standard geological column is not entirely clear but find that, in general, the arrangement as given in the column is good for most of the world.

Although the lateral persistence of facies is an asset for stratigraphical work, explanations of the phenomenon were rarely attempted by geologists in the Lyellian tradition. Ager (1993) considered the subject worthy of particular comment. After reviewing numerous examples of widespread facies, he commented:

> So, as one goes down the stratigraphical column, if one leaves behind the spectacles of the specialist and looks about one with the wondering eyes of a child, one never ceases to be amazed at the diversity and yet the uniformity of it all. No doubt my readers will have bigger and better examples of the persistence of facies which so fascinates me in this chapter, but I write as far as possible from my own experience (p. 21).
>
> Those who are fascinated by the minutiae of stratigraphical correlation may be horrified at my generalisations. But I find myself left with what may be called the first main proposition of this book: *At certain times in earth history, particular types of sedimentary environment were prevalent over vast areas of the earth's surface*. This may be called the 'Phenomenon of the Persistence of Facies'" (p. 23, emphasis in original).

There is no satisfactory rationale for these features within the conventional uniformitarian paradigm. Plate tectonics theory offers some ideas, but they are not very convincing, because most modern geological interpretations are built on modern day analogues, where persistence of facies is not a feature. Only diluvialism operates on a genuinely global scale.

Global patterns of sedimentation are found particularly in Precambrian and Paleozoic series of rocks. For example, the Redwall of the Grand Canyon is continued as the Madison to the north. Equivalent beds are found in the Canadian Rockies (Rundle Formation), in New York (Niagara Falls flows over an equivalent bed). It is found in Europe and makes up the upper part of Mount Everest. The phenomenon is less pronounced, but still a feature of Mesozoic rocks. Numerous examples are provided by Ager (1993) and Morton (1984). The key point to note here is that these widespread lithologies can only be regarded as data if the geological column concept is viable, and that meaningful correlations can be made.

Paleocurrent directions are being analyzed by Chadwick (1993). For Paleozoic sediments and, to a lesser degree, Mesozoic sediments, the paleocurrents trend generally in the same direction all the way across the North American continent. This simply does not fit with gradual accumulation of sediments over millions of years and, like the persistence of facies data, points to the inappropriateness of uniformitarianism as a geological tool.

Principles for establishing stratigraphical correlation

Several principles for correlating strata have emerged thus far in this chapter. These can be summarized as follows:

1. Field relations (the law of superposition)
2. Lithology (often characterized by the persistence of facies)
3. Associated fossil content (the law of faunal succession)

Two additional criteria for correlation can be added.

4. Degree of diagenetic change (metamorphism, lithification, alteration)
5. Event Horizons (volcanic ash fall layers, impact crater debris, earthquake deformation horizons).

These latter two criteria are considered further below.

Degree of diagenetic change

Rocks show many kinds of alteration fabrics, ranging from the mode of lithification to the presence of metamorphic minerals. Whilst these changes can sometimes be complex, at other times, they can be very useful for assisting with stratigraphical correlation.

The early division of rocks into Primary, Secondary and Tertiary illustrates this. The definitions of these terms were never exact, but they did communicate something of significance. The rocks found towards the base of any sequence are more crystalline, harder and often lack fossils or sedimentary structures. Above them are a series of strongly lithified rocks that often contained fossils and sedimentary structures. Above the Chalk (which was regarded as the top of the Secondary formations) are the Tertiary strata, much of which is poorly lithified. This perception of the crystalline/lithification state of the strata has been used to assist the process of correlation. Although today's terminology is different, the same concept continues to be valid.

As an example of how such data can be used by diluvialists, consider the Rioja region of Spain (Garton, 1997). It was found that all the lower strata, from Precambrian to Lower Carboniferous, had been metamorphosed in the Hercynian mountain building episode. The rocks contain low grade metamorphic minerals that give them a shiny appearance. A pronounced unconformity separates the metamorphosed rocks from the overlying strata. Thus, mountain building was followed by erosion before the deposition of Mesozoic strata (which would have been classified as a Secondary formation by Werner). Tertiary deposits contain conglomerates made up of cobbles of Jurassic, Triassic and Cambrian rocks (recognizable by their lithologies and degree of metamorphism). They provide evidence that these strata were uplifted again and eroded to form new depositional units. Garton (1997) writes:

> These mountain building events prove that, at least in Rioja, the major divisions of geological

> strata—Precambrian, Palaeozoic, Mesozoic, and Tertiary—are distinct, are separated in time and appear in that order. There has been no appeal to fossil or radiometric evidence (p. 17).

In the Grand Canyon of the Colorado River, most of the Precambrian and Paleozoic sediments are crystalline and well-indurated. They hold up as cliff formers in the steep walls of the Canyon. The Mesozoic and Tertiary beds mostly have been eroded away. Because the Mesozoic and Tertiary sediments are usually softer, topography involving these sediments is often that of badlands, not steep-walled canyons. Many such examples can be given. Thick beds of soft shales, such as the Mancos, Pierre, and Cody in North American Rocky Mountains, are less obvious in the Paleozoic.

An example analogous to the Rioja case study given above comes from an unusual product of erosion. The Madison Mountain Range in Montana includes the unique Sphinx Mountain (Figure 8) that is composed entirely of conglomerate rocks of pebble to cobble size. With an elevation of nearly 4,000

Figure 8. Sphinx Mountain, Montana. This mountain is entirely composed of breccia and conglomerate in reverse stratigraphic order. See text for more detail.

m, it is one of the highest mountains in the range. The conglomerate is composed of Paleozoic and Mesozoic rock. This rock is said to have washed into a basin from an adjacent mountain range to the west. After the sediments were washed into the basin, the basin was uplifted and/or the adjacent mountain range was lowered. Subsequent erosion has left Sphinx Mountain standing tall. An unusual feature of the stratigraphy of Sphinx Mountain is that the lower part of the mountain is composed of softer, smaller pebbles eroded from "younger" sediments, whereas, the upper part is composed of larger, harder "older" rock.

How can this be explained? The adjacent mountain range to the west contains Paleozoic and Mesozoic stratigraphy in correct order. Erosion took off the tops of the mountains (Mesozoic sediments) in this range and dumped them into the basin. When erosion cut down to the lower "older" Paleozoic sediments, they were next dumped into the basin on top of the "younger" rocks. Thus Sphinx Mountain has a reverse order of "younger" rocks below the "older" but the explanation for this "mystery" is based on the erosion of strata that agree with the order of the geological column.

Event Horizons

Event horizons can be very useful in studies of local areas, and are exceptionally useful for establishing correlations over much wider areas. The principle is that a specific event (e.g., a volcanic eruption [Figure 9], a meteor or comet, or an earthquake) leaves its mark in the rocks. Since the event occurs over a very

Figure 9. The light-colored sediments, in southern Alberta, Canada, are bentonite (weathered volcanic ash) that no doubt came from somewhere to the west.

short timescale, an event horizon becomes a tool for stratigraphical correlation.

Some coal seams in North America and Europe contain thin partings only a few centimeters thick (tonsteins) that can be traced great distances—from England and throughout much of Western Europe. Obviously these tonsteins make excellent markers that aid correlation. The most reasonable explanation for tonsteins is that volcanic ash or other sediments were laid down over a wide area in the midst of the sedimentation that produced the coal seams.

The volcanic eruption of Mount Mazama, the remains of the mountain now containing Crater Lake, (Figure 10) spewed ash thousands of square km in the Pacific Northwest and Southwest Canada. This eruption, considered to have been only a few thousand years ago, has left a white band only a few centimeters or less thick in peat bogs and near the surface of Quaternary sediments.

The best example of a volcanic ash layer comes from the Ordovician (Bergström et al. 2004). In North America, the Deicke and Millbrig beds are over a meter in thickness on the east coast, thinning to the west. In Northern Europe, the Kinnekulle bed can exceed 2 m and thins to the east. These beds can be traced over millions of km^2. Their stratigraphic position on both sides of the Atlantic makes them equivalent, so there has been great interest in analyzing them chemically to see if they can be traced to the same eruption. Although some differences can be found, there are also parts that are indistinguishable

Figure 10. Crater Lake with a volcanic cone that arose in the lake.

geochemically. If the ash horizons are confirmed as belonging to the same event, they provide an impressive example of correlation between continents.

The geological literature is replete with information about the end-Cretaceous impact event, predicted in 1980 by Alvarez and co-authors. After an extended search, the diagnostic features left by an impact were found at the Cretaceous/Tertiary boundary (K-T) on the Yucatan Peninsula in Mexico. The Chicxulub crater has been the subject of research (and controversy) since then. A creationist review of the literature up to 1996 was provided by Fritzche (1998), who made these comments on the value of the stratigraphical work prior to the impact hypothesis:

> The way scientists searched for K-T sections to check the Alvarez proposal is fine evidence, they only had to look at the geological maps and descriptions. So, the K-T boundary with all its implications is a fine illustration of what stratigraphy is good for, how it "works" when scientists are challenged to test for a global event (p. 246).

Other examples exist where the event is not so sudden. Chalk is such a distinctive rock in composition and in most parts of the world, a distinctive part of local geological columns. Garton (1990, p. 5) draws attention to this and suggests "we must allow for an event which has caused all this chalk to form." An even better known case is the formation of the Cambrian unconformity, which is almost global in extent and which separates Precambrian strata, generally characterized by an absence of macrofossils, with the Cambrian, which is characterized by an explosion of hard-bodied marine animals. This unconformity at the base of the Cambrian is well-known from exposures in the Grand Canyon, but the same characteristics are found all over the world. An unusual, more localized case concerns disturbed bedding linked to the rapid flooding of the Weissliegend sandstones (Glennie and Buller, 1983).

Defenses of the Geological Column concept

The first modern diluvialist to defend the concept of the geological column was Clark (1946). He consciously sought to correct some of the demolition work undertaken by George McCready Price, who had once been Clark's teacher. Clark found himself advancing different views because, in his opinion, the field evidences demanded it (an overview of the differences between Clark and Price is given by Numbers, 1993, pp. 123–129). Clark addressed the development of American stratigraphy, their correlation with the rocks of England, and elaborated on the principles by which rocks may be correlated. He wrote:

> The diluvialist may not agree with geologists in interpreting the order of the strata as evidence of long ages of evolution, but this objection to the evolutionary theory should not lead them to close their eyes to the fact that the rocks do lie in systematic order. The detailed correlation that has been made is too exact and clear to make possible a successful denial of a very definite and regular order in the fossil zones (Clark, 1946, p. 52).

Clark then elaborated on ecological zonation as his explanation of the fossil successions.

Many years later, Ritland (1981; 1982) reviewed the history of the development of the geological column and found that most of the geologists concerned were people who were not driven by evolutionary theory. He concluded:

> Completely apart from any merits or weaknesses, the geologic column is the result of an attempt by conscientious scientists to construct to the best of their ability a classification of rock strata that would account for the phenomena encountered in the crust of the earth (Ritland, 1982, p. 47).

Coffin (1983) explains the nature of the stratification of the Earth's rocks in Chapter 6 of his book. This considers rocks found in the United States and how they are correlated. He summarized:

> ...the rock beds of the earth are arranged like the layers of a gigantic cake. Furthermore, their order generally is consistent and predictable. A substantial part of the geologic column is present and in correct order in some areas. In many other locations much is missing, but those few strata that are there lie one above another in correct sequence unless mountain building or faulting has disturbed them (p. 68).

Watts (1985) set out to clarify the distinction between the geological column as a genuine sequence of rock strata and the time spans linked to the column. He identified the common principles for correlating strata and then considered several matters arising. In particular, he noted how inter-continental correlations could be made.

Garton (1990) developed the concept of an onion-skin Earth, where local geological columns around the world show the same order in specific rock units. He chose four distinctive rock types to make the point, listed here so that (1) lies above (2), etc:

1. Rocks showing evidence for an ice age;
2. Chalk formation, after which the dinosaurs die out;
3. Rocks containing dinosaur bones and tracks;
4. Rocks linked to the "Age of coal formation."

Garton also addressed claims that there are numerous places on the Earth where the sequence is contradicted and points out that these always come from disturbed areas known as mountain belts. Out of sequence layers are generated after deposition by the process of overthrusting.

In a statement about the geologic record (Snelling et al., 1996), an international group of diluvialists distinguished concepts that are often conflated within the geologic column: particularly the lithostratigraphic column, the biostratigraphic column and the chronostratigraphic column. The general validity of the first two elements was affirmed and considered to be a sound basis for future research in creationist geology. This statement also affirmed "that the starting point for our geological thinking is the historical record in the book of Genesis, which we accept at its face value."

Robinson (1997) has contributed the most thorough recent analysis of the issues, starting with a local column and discussing the basis on which regional, national and international correlations can be made. He then examines and answers the three most-weighty objections that have been made to the geological column concept. He explains that acceptance of the geological column does not mean that we adopt everything that others might affirm. In particular, he identifies three things *not* entailed by the geological column:

1. It does not entail acceptance of the multi-million-year periods.
2. A correlation based purely on biostratigraphical criteria—say, between strata on different continents—does not establish absolute synchronicity.
3. The geological column does not represent a systemization of perfect synchronisms.

He concluded (pp. 24–25):

> The 'geological column' is an aid to understanding, not an obstacle. Collations of data which present the global picture as it changes over time—such as sea-level curves, diagrams showing the patterns of fossil groups in the record, major extinction events, the incidence of continental flood basalts—are all made possible by the correlations which the geological column embodies, and all cry out for explanations which a global flood is uniquely fitted to provide. The Flood cannot be effectively argued for geologically without the column. It is the common ground with old-earth ge-

ology—and the common language—which allows interpretations to be compared in meaningful dialogue.

Brand (1997) has a section on the geologic column in Chapter 13 of his book. After reviewing the stratigraphy of some locations in the United States, Brand wrote (p. 240):

> The sequence of fossils in the rocks apparently is real. Whether we prefer catastrophic geology or conventional geology, the geological column is still a valid description of nature's history book.

Conclusions

A case has been made in this chapter for a positive approach to the geological column concept. We have argued that the column is a servant, not a master, and there is no necessity to link it to any particular timescale. Arguments concerning the stratigraphic position of a particular stratum are not unusual in academic literature. However, these debates usually relate to relatively minor changes from one Period or Epoch to an adjacent one. Rarely is the controversy over a major change from one Era to another for instance.

As a servant, the column must not be allowed to prejudge field data. It is not good practice to use fossils exclusively for stratigraphical correlation, for then the column becomes the master and is open to the charge of a "vicious circle of reasoning" (Whitcomb and Morris, 1961, p. 203). Out-of-order fossils do occur occasionally. These are often ignored by paleontologists or accounted for by reworking of older or the intrusion of younger sediments. Other explanations sometimes seen in the literature are changes to the habitats of fossil biota as compared to similar modern specimens (e.g., marine to fresh water, Coffin, 1971) or assumptions that a fossil is a different species or genus even though it is identical to modern representatives. Diluvialists should be champions of careful analysis of such anomalies, evidencing a willingness to challenge the current "orthodoxy."

Nevertheless, despite these possibilities for stratigraphical revision, the major features of the geological column hold up well as determined by field observations of the fossil content, petrology, color, hardness, and the relationship of beds to each other. As such, the column becomes a tool to use and a key ingredient assisting our goal of reinterpreting historical geology within the framework of biblical history.

References

CRSQ—*Creation Research Society Quarterly*

Ager, D. 1993. *The nature of the stratigraphical record,* third edition. John Wiley & Sons Ltd., Chichester.

Bergström, S.M. Huff, W.D. Saltzman, D.R. Kolata, and S.A. Leslie. 2004. The greatest volcanic ash falls in the Phanerozoic: trans-Atlantic relations of the Ordovician Millbrig and Kinnekulle K-Bentonites. *The Sedimentary Record* 2(4):4–8.

Brand, L. 1997. *Faith, reason and Earth history*. Andrews University Press, Berrien Springs, MI.

Beutner E.C. and G.P. Gerbi. 2005. Catastrophic emplacement of the Heart Mountain block slide, Wyoming and Montana, USA. *Geological Society of America Bulletin* 117(5):724–735.

Chadwick, A.V. 1993. Megatrends in North American paleocurrents. *Society of Economic Paleontologists and Mineralogists Meeting Abstracts* 8:58. See also: http://geology.swau.edu/paleocurrents_1.html (accessed December 2005)

Clark, H.W. 1946. *The new diluvialism*. Science Publications, CA.

Coffin, H.G. 1971. A paleoecological misinterpretation. In Lammerts, W.E. (editor). *Scientific studies in special creation*. pp. 165–168. Presbyterian and Reformed Publishing Company, Philadelphia, PA.

———. 1983. *Origin by Design.* Review and Herald Publishing Association, WA.

Cutler, A. 2003. *The seashell on the mountain top*. Dutton, New York.

Fritzche, T. 1998. The impact at the Cretaceous/Tertiary

boundary. In Walsh, R.E. (editor). *Proceedings of the Fourth International Conference on Creationism.* pp. 241–251. Creation Science Fellowship, Pittsburgh, PA.

Garton, M. 1990. Rocks and Scripture: investigating the earth's history. *Origins (Journal of the Biblical Creation Society)* 3(9):4–9.

———. 1997. A Spanish weekend. *Origins (Journal of the Biblical Creation Society)* 22:11–24.

Glennie, K.W. and A.T. Buller. 1983. The Permian Weissliegend of NW Europe: the partial deformation of aolian dune sands caused by the Zechstein transgression. *Sedimentary Geology* 35:43–81.

Gould, S.J. 1987. *Time's arrow, time's cycle.* Harvard University Press, Cambridge, MA.

Miller, H. 1852. Geological evidences in favour of revealed religion. In Miller, H. 1906. *The Old Red Sandstone.* pp. 273-283. J.M. Dent & Sons Ltd., London.

Morton, G.R. 1984. Global, continental, and regional sedimentation systems and their implications. *CRSQ* 21(1):23–33.

Numbers, R.L. 1993. *The creationists.* University of California Press, CA.

Ritland, R. 1981. Historical development of the current understanding of the geologic column: part I. *Origins* 8(2):59–76.

———. 1982. Historical development of the current understanding of the geologic column: part II. *Origins* 9(1):28–50.

Robinson, S.J. 1997. The geological column: a concept foundational to Flood geology. *Origins, Journal of the Biblical Creation Society* 23:14–30.

Rupke, N.A. 1983. *The great chain of history.* Clarendon Press, Oxford.

Snelling, A.A., J. Scheven, P. Garner, M. Ernst, S.A. Austin, M. Garton, E. Scheven, K.P. Wise, and D. Tyler. 1996. The geological record. *Creation Ex Nihilo Technical Journal* 10(3):333–334. Also in *Origins, Journal of the Biblical Creation Society*, 1997, 22:28–29.

Watts, D. 1985. Rock strata and the geological column. *Biblical Creation* 7(20):59–65.

Whitcomb, J.C., Jr. and H.M. Morris. 1961. *The Genesis Flood.* Presbyterian and Reformed Publishing Company, Philadelphia, PA.

Authors

David J. Tyler has a background in physics (BSc and MSc) and a career that has led him into management (PhD). He has 8 years experience in applied research plus other industrial experience. As an academic, he is involved in teaching and research. His first paper in the CRSQ was in 1979 and he has been actively writing and speaking since then. He coordinates a group in the UK that is actively seeking to develop geological science built on biblical foundations.

Harold Coffin earned a PhD in Zoology and spent much of his career with the Geoscience Research Institute, retiring in 1991. He has published six books and numerous papers on paleobotany and sedimentology. He is known for his landmark studies of the Yellowstone fossil forests. As a result, the park no longer promotes the fossil trees as in-situ deposits.

Forum Discussion

1. The term, "column" is used ambiguously, switching back and forth between the empirical rock record and its features, and the set of interpreted global correlative synchronous time periods. (Reed, Klevberg, Froede)

We see the empirical rock record as a set of local columns. We advocate a "bottom-up" approach to correlation and regard the global Geological Column as the product of empirical data collection and analysis. In our chapter, the word "column" is usually accompanied by another word that makes clear our meaning. Thus, "local column" refers to the empirical rock record. In other cases, the context is intended to establish the meaning. Hence we do not think we have been ambiguous.

2. Examples of inherent order in the rocks and features such as regional marker beds do not demand an interpretive framework of global correlative synchronous time periods. (Reed, Klevberg, Froede)

Correct. Nevertheless, these observations provide "handles" for correlation which, in our judgment, lead to global correlation. We have not presented global correlation as an "interpretive framework" but as a research outcome.

3. When looking at a section like the Grand Canyon, do missing sections of the column reflect missing time, missing rock, or a gap meaningless to diluvialists? If missing rock or time, how much is missing and how would we know? If meaningless, does not that decrease the relevance of the column to diluvialists? (Reed, Klevberg, Froede)

These questions cannot be answered in the abstract. It is the goal of fieldwork to gather relevant evidences that advance understanding of the geological history of an area. We would not suggest that any field evidences are "meaningless to diluvialists." Considering the Grand Canyon, it is a worthy goal for diluvialists to develop a geological history of the region that does justice to the field evidences. How this can be done without addressing the issue of correlation across regional boundaries is a mystery to us.

4. How can diluvialists consider the chronostratigraphic column useful or valid when nearly ninety percent of it (the Precambrian) is defined by the geochronology that we all reject? (Reed, Klevberg, Froede)

In the Precambrian, there are meaningful lithological differences that facilitate mapping of local columns. Continental cratonic areas have been identified by mapping, together with the division between Achaean and Proterozoic sequences. In the Upper Proterozoic, there are distinctive biostratigraphical markers that supplement the lithological data. This may not be enough to satisfy advocates of the "chronostratigraphic column," but it gives diluvialists plenty to work with.

5. In "Connecting Local Columns" you make the statement in the first sentence that certain persistent lithologies have characteristic fossils. There

have been many statements in the literature that lithologies and fossils, on a global scale, do not correlate. How do you reconcile these? (Oard)

Overall, we wish to affirm our conclusion that persistent lithologies do have characteristic fossils. We recognize that there are exceptions, but these do not reduce the strength of our argument. We do not accept that there are "many statements in the literature that lithologies and fossils, on a global scale, do not correlate."

6. Granted that the lithostratigraphy of England may be well behaved. In "Connecting Local Columns" (paragraph 1) you state that the stratigraphy of England is close to that of France and Germany. How close? What about other parts of Europe? (Oard)

The Mesozoic and Cenozoic rocks of England can be traced across the Channel to the Paris Basin. The Paleozoic rocks of SW England have links with the Channel Islands and Brittany. The Highlands of Scotland have many links with Scandinavia. Paleozoic rocks in England appear in Germany and beyond. One of the biggest differences is associated with the Permian, where continental Europe has extensive limestone beds sandwiched between red beds. These particular limestones are absent in England. The closeness can be assessed by looking at the 1821 lithological column (largely based on England, Germany and the Paris Basin) and subsequent work. The main elements are confirmed, there are some additions, but mostly it is fine tuning.

7. You state that overthrusts and reworking solve contradictions to the predicted order of the geological column. Granted that Heart Mountain is a detachment fault in which the upper block slide downhill and broke up, what other evidence shows that large-scale overthrusts and reworking can explain examples of out-of-order strata (see notes on the Carpathians and reversed metamorphic gradients in Chapter 6)? (Oard)

Diluvialists have been seriously misled about overthrusts. Normal and reverse faulting cannot just be scaled up to predict what overthrusting should be like. One of us (DT) first realized this when inspecting the Moine overthrust in Scotland. This has metamorphic gneiss thrust over Cambrian sedimentary rocks. The major thrust planes are remarkably sharp. It is possible to put a finger on these planes. There is limited deformation in the gneiss, with some mylonization, and some shearing and fragmentation of the Cambrian rocks. This rapidly diminishes away from the thrust plane. There is no evidence of fault breccias and the other diagnostic features that diluvialists claimed should be present if this structure was an overthrust. Elsewhere, where there are extensive duplex structures, the contacts are sharp, with no macroscopic signs of deformation. These observations have since been reinforced by fieldwork in other areas (such as the European Alps and in Spain) where the evidences of significance are mylonites at the fault plane, evidences of pressurized water in the thrust plane, and mapping of thrust blocks.

five

Recolonization and the Mabbul

David J. Tyler

Abstract

This chapter provides an overview of the recolonization approach to understanding biblical teaching on the Flood and the geological record. Recolonization accepts the geological column and seeks to understand the fossil and lithological successions within a framework of biblical history with short timescales. The "mabbul" destruction in the time of Noah is shown to have implications for geology in the past and for Earth in the future. After explaining the foundations of the approach, the chapter considers the key stages in the post-Flood restoration of the planet and relates this to geological evidences.

Biblical Foundations

The Flood Executed God's Judgment on all Humanity Except Those in the Ark

The Flood came as a judgment upon the whole of humanity because it had turned completely away from God (Genesis 6:5).[1] God was "grieved that He had made man on the earth" (Genesis 6:6). It was not the case that some sections of the community were better than others. The judgment was to be universal. Out of the whole human race, only Noah's household was saved. No others were allowed to escape. The situation is analogous to the future before the Second Coming of Christ (Matthew 24:38, 39). All of humanity was involved in a day of judgment in the time of Noah, and all will

[1] New International Version throughout chapter.

be involved in the Day of Judgment when Christ returns. This universality of judgment necessarily leads to a global flood because of humans' migratory tendencies.

There is a huge credibility gap in holding to a local flood model and at the same time affirming the anthropological universality of the Flood.

The Flood was a Sudden Destruction

We need to be able to talk about the Flood like Jesus Christ did. In His preaching, there were repeated warnings of sudden destruction. In Matthew 24:38–41 and Luke 17:26–35, Jesus refers to the Flood beginning suddenly. He drew this analogy between the judgments in the days of Noah, on Sodom and His future return. Each judgment is presented as sudden:

1. The people carried on as usual until the day Noah entered the Ark.
2. The day Lot left Sodom was the same day that the city was destroyed by a downpour of fire and sulphur.
3. On the day Christ returns, people may be in the home or they may be in the field, but His coming will bring their activity to an abrupt conclusion and usher in the end of the age.

On one particular day (Genesis 7:11), all (not one!) springs of the great deep burst forth, the floodgates of heaven were opened and torrential rain fell for 40 days. On that day, the eight people and the animals entered the Ark. Delay, even for a day, was not an option. They needed its protection. It was not long before the Ark was afloat, preserving those inside while all outside "that had the breath of life in its nostrils died" (Genesis 7:22). Did any survive outside of the Ark after the first day? The bursting forth of all the springs is an indication that the entire Earth was awash with surging water on the very first day. If anything did survive, it could not have been much longer than a day since the inundation was so overwhelming. The parallel Jesus drew with Sodom and with His own return indicates that the destruction of life was over quickly.

This means that much diluvialist thinking is wide of the mark, because many want there to be a period of *40 days* when the waters encroached onto the land, allowing animals to breed and make nests while they retreated from the rising water (over sediments already laid down by floodwater). This scenario does not do justice to the biblical record. Furthermore, local flood approaches are hopelessly deficient because there is always a way of escape from a local flood. From the Genesis Flood, there was neither time nor place to escape.

The Flood Destroyed the Land with its Non-Human Inhabitants as well as Humanity

The reason for Earth's destruction is found in the relationship between Adam and Eve and the world God gave them to rule (Genesis 1:28). They were given dominion over everything (Genesis 3:17), so after the Fall, God not only punished Adam and Eve, he also put the marks of his anger on the creation itself. This is the Edenic curse—what Paul described as the creation "groaning as in the pains of childbirth" (Romans 8:22). This creation experienced judgment when the descendants of Adam were judged. It is the same creation that will one day undergo the fires of judgment. It will again be destroyed—only to be transformed and "liberated from its bondage to decay" (Romans 8:21). On that day, Earth will enter into the redemption that Christ has purchased, and there will be the New Heaven and the New Earth.

The Recolonization approach incorporates the understanding that God destroyed the antediluvian land surface over which mankind had dominion—a point important both theologically and historically. Attempts to diminish this aspect of revealed truth undermine the coherence of Scripture as a whole.

The Flood was a type and a foreshadowing of what will happen to the universe when Christ returns (II Peter 3:5–13). While the Flood was a lesser destruction, it was of enormous significance in Earth's history. A local flood will not do.

The Mabbul

In Genesis, the word "mabbul" is used of the destruction. It applies to the first 40 days of the time when Noah was on the Ark. The mabbul is the "cataclysm." It was overwhelming. After these 40 days, the springs of the great deep closed, the floodgates of heaven stopped, and the torrential rain ceased, signifying a cessation of destruction. After 150 days, God remembered Noah (Genesis 8:1) and the waters started to recede. Restoration was underway (c.f., Robinson, 1998; 2000a).

Scientific Methodology

Two issues are noted here that are relevant to biblically-based science.

a. The methodology of historical sciences (historical geology, evolutionary biology, archaeology, forensic science) differs in numerous areas from the methodologies used in empirical sciences (Cleland 2001; Tyler 2002). In particular, historical sciences cannot carry out experiments that relate directly to past historical events, which are unique. This is a limitation that provides constraints, but does not rule out genuine science being undertaken. However, the shoehorning of data to fit preconceived theories is always to be deplored—whether it is done by uniformitarians or catastrophists.

b. The phenomenon described as "God of the gaps" is not encouraged. Miracles are historical events, but it is inappropriate to invoke miracles just to make sense of the data.

Geological Evidence for a Global Flood

Many Flood geologists have put great significance on the global Cambrian unconformity and the Cambrian rocks being the lowest strata containing hard-bodied fossils. Some have thought that the unconformity surface was the base of the Flood erosion (Baumgardner, 1990) or at least not far below it (Austin and Wise, 1994). However, there has never been uniformity of opinion on these matters, partly because Whitcomb and Morris (1961) were prepared to consider some Proterozoic rocks as Flood deposits and because they pointed out that "there are a number of important localities where the transition from Proterozoic to Cambrian is not marked by an unconformity of a physical nature but rather by a disconformity, the only evidence of the change being fossiliferous" (p. 231). A major problem for locating the base of the Flood at or below the base of the Cambrian is that the evidences for catastrophism below the unconformity are often greater than the evidences of catastrophism above it!

The macroscopic scale of Precambrian rocks generally, and Archean rocks in particular, is pervasive. As an example, visitors to Snowdonia National Park in North Wales can inspect the Padarn Tuff near Llanberis. This Precambrian welded volcanic ash fills a vast channel and is at least one (if not two) km thick. Significantly, there are no discernible bedding plane horizons to mark successive flows or surges of ash. The whole unit is welded together without distinct breaks. The volcanic activity producing this deposit is utterly unlike anything going on today; it was an extraordinarily catastrophic event. To appreciate the Precambrian, it is necessary to abandon mental models based on present-day processes.

The Flood released extraordinary destructive power. The "springs of the great deep" which "burst forth" is hardly a technical description. The words need interpreting in terms of geological activity. They are understood here in terms of global crustal fracturing, leading to catastrophic surging of water followed by magma released from within the Earth. Accompanying this was the outpouring from the windows of heaven (which may be linked to bolide impacts) and torrential rain. These ingredients establish some benchmark principles for diluvialism:

1. The scenario is not one of the sea rising gradually to flood the land. This option fails to capture the violence and the immediacy of the occasion. Rather, it does not really matter what is land and what is shelf sea—the cataclysm affected both. Because the continents and shelf seas are somewhat elevated on Earth's surface, the mabbul would have overwhelmed everything belonging to the antediluvian world, carrying all into the ocean basins. Following Robinson (1996), the recolonization model advocates subterranean water bursting from below the continents, immediately scouring everything *from* the land *into* the sea. Therefore the Flood has little to do with gradually rising water levels (c.f., Whitcomb and Morris 1961; Wise 2002; Brand 1997) and everything to do with cataclysmic judgment cutting people down and scouring them off the land.

2. With fracturing of the crust and release of its waters, decompression effects would have been significant, and hot rocks at depth would have melted and flowed upwards. This is envisaged as an avalanche process. Since the surface was covered with water, this would have resulted in pillow lavas, peperites and hyaloclastic deposits, in addition to extensive igneous activity and massive sedimentation. All the sediments coming from the antediluvian world would be mingled with the products of igneous activity, accompanied by extensive metamorphism, and this had the effect of obliterating anything that might potentially have become a fossil. In other words, the world that then was, was destroyed (II Peter 3:6). For more on peperites, see Walker (2004), noting that these textures are widely documented in the Precambrian.

The Recolonization approach involves the destruction of the land of the antediluvian world. This requires a global destruction horizon at the same time as the whole Earth was covered in water. Such a horizon is discernible in the rock record—with the crustal reworking being located in the Hadean (Kamber et al., 2005, 276–277; Harrison et al., 2005, 1947). These authors commented as follows.

> In contrast to the antiquity of the Martian and lunar crust, exceedingly few fragments of terrestrial crust from the first billion years have survived. While rocks younger than 4.0 Ga have been discovered on several cratons, the apparent complete absence of bona fide >4 Ga (Hadean; here informally defined as older than 4 Ga) terrestrial rocks is both surprising and enigmatic.... In view of such unambiguous evidence for their former existence, the apparent lack of preserved Hadean (i.e., >4.0 Ga) crustal rocks is therefore a fundamental observation that must be explained by all models of silicate differentiation in the early Earth (Kamber et al., 2005).
>
> The long-favored paradigm for the development of continental crust is one of progressive growth beginning at ~4 Ga.... These results support the view that continental crust had formed by 4.4–4.5 Ga and was rapidly recycled into the mantle (Harrison et al., 2005).

The significance of this paradigm change should not be lost: Earth did not start to accumulate a continental crust after the Hadean, it had a continental crust at the beginning of the Hadean but it was "rapidly recycled" and the reforming process began in the Archean. This scenario is superior to that of Rosing et al (2006) who invoke photosynthetic microbial life as the trigger for the formation of crustal materials after about 4 Ga.

The mabbul corresponds to the Hadean and possibly much of the Archean, during which Earth's crust (to depths of at least 10 km) was reworked. One of the few diluvialists to recognize the Archean's importance (apart from the Recolonizers) is Hunter (1996, p. 349).

> The five detailed stratigraphic sections through Archaean strata... show that the lower Archaean strata are comprised dominantly of extrusive volcanics with basal conglomerates and some quartzites, cherts, graphitic shales and limestones. The volcanic sequences are usually terminated abruptly at their top by a sudden transition to sedimentary strata. This transition may have resulted from sudden cessation of the major activity of 'the fountains' at the end of Day 40 of the Flood.

The Archean high-grade metamorphics and igneous bodies are massive, and require thinking on a grand scale to understand their formation. No body fossils are present—only biochemical traces of life. There is evidence that water temperatures were high: 55–85°C.

> Oxygen isotope data combined with the results of geological and sedimentological studies demonstrate that enclaves of syn-sedimentary to very early diagenetic cherts are widely preserved in the 3.5–3.2 Ga Swaziland Supergroup, Barberton greenstone belt, South Africa. The low ^{18}O values of these cherts indicate extremely high ocean temperatures of 55–85° C (Knauth and Lowe, 2003. p. 566).

There is very limited data on water temperatures, but the prediction of the Recolonization approach is that the South African data is representative of much of the Archean. This is borne out by Knauth's (2005, p. 53) analysis:

> Oxygen isotope data for early diagenetic cherts indicate surface temperatures on the order of 55–85° C throughout the Archean.

In summary, the Hadean/Archean was a time of "meltdown" for the Earth. While conventional geology has viewed this time as the initial establishment of continental crust, it is now recognized by a growing number of geologists that the early Earth was water-covered, that the Hadean crust did once exist and was destroyed, and that new crust was formed in the Archean and Proterozoic. Recolonization provides a reason and a mechanism for the destruction of Earth's former continents. The Hadean is interpreted as the horizon of destruction followed by the remaking of Earth's land surface. The geological data therefore complement the biblical record of the mabbul destruction, when God destroyed the Earth along with all air-breathers, save those in the Ark.

After the destruction of the mabbul, the first continental environments to be recolonized were marine, as the surface of what was to become land was still covered with water. Only later, as the land emerged above the sea, could there be terrestrial recolonization.

An Oxygenated Atmosphere

Contemporary thinking about the Precambrian atmosphere is unduly influenced by a commitment to abiogenesis. That paradigm requires the Earth to have had a reducing atmosphere at the outset, evolving to a neutral atmosphere and then, with a developing biosphere, slowly increasing in oxygen content to reach the modern mix. The supporting evidence for reducing/neutral atmosphere, however, comes primarily from sediments in marine environments—where even today reducing environments can be found!

To support the claim of a reducing atmosphere, it is necessary to look at paleosols. Various claims were made for these, but a major reexamination of the paleosol evidence was made by Ohmoto in 1996. He concluded that a reducing or neutral atmosphere was not consistent with the data. He argued that the minimum oxygen pressure for the 3.0–2.2 Ga period (no earlier paleosol data is available) has to be about 1.5% present atmospheric levels (c.f. Tyler, 1996a). The significance of the results is that photosynthesizing organisms must have been present, at least back to 3.0 Ga, to sustain oxygen levels. The problem lies in the lack of evidence for these organisms. Ohmoto (1996, pp. 1135) certainly was thinking along these lines: "Terrestrial biomass on the early continents may have been more extensive than previously recognized." Thus, conflicting ideas exist about oxygen in the Archean atmosphere. From a Recolonization perspective, the oxygen had a pre-mabbul origin, and oxygen producers may or may not have been present, depending on the state of re-establishment of the post-Flood oxygen cycle.

Molecular markers have been found that make a link to photosynthesis. Some are found in Early Archean sediments. Rosing and Frei (2004, p. 243) concluded:

> The high biomass productivity of planktonic organisms, the strongly ^{13}C-depleted carbon isotopic signature and the evidence for the presence of oxidized aquatic environments all suggest that oxygenic photosynthesis had developed before 3700 Ma.

Some geologists regard these results as firm, but others are skeptical. If they do provide evidence for oxygen producers, the recolonization approach interprets them as pioneer species in the restoration process.

Archean oil finds also support the concept of a more extensive biomass. Dutkiewicz et al. (1998) have reported oil in fluid inclusions from six Precambrian cratons dating from ~2,450 Ma to ~3,000 Ma (with metamorphic ages only slightly less than these). The authors find that the oil was emplaced before metamorphism sealed the voids, making some of it of Archean age. Some of the interesting implications of this are pointed out in the article, with selected quotations below.

> The discovery of oil preserved in Archaean fluid inclusions ... confirms the contention that hydrocarbon generation by thermal maturation of biogenic kerogen was extensive in Archaean sedimentary basins. This is somewhat at odds with conventional wisdom about petroleum distribution, which regards Archaean terrains as essentially unpetroliferous (p. 887).
>
> As hydrocarbon generation and migration were evidently widespread phenomena in Archaean and early Palaeoproterozoic basins, global burial of organic carbon must have been substantial during these early stages of the Earth's history. Indeed, isotopic mass balances indicate that at least 12% of all buried Archaean carbon was organic.... Indeed, as nutrient levels are unlikely to have been strongly limiting in Achaean water bodies, with exposed continents supplying phosphate and nitrogen-fixing cyanobacteria probably present, the Archaean aquatic biota may well have been as productive as its modern counterpart (p. 887).

Although there are continuing claims in the literature for very low oxygen levels in the Proterozoic, there are plenty of evidences for significant oxygen in the atmosphere at that time. Melezhik et al. (2005) review these and refer to "the Great Oxidation Event (GOE)" which they date in the period 2,450 –2,320 Ma. The enigma for them is that "the cause of the GOE remains unresolved..." (p. 6).

Marine Recolonization

One possible interpretation of all this evidence that the marine Archean and Proterozoic was teeming with bacterial and algal life. This implies that signs of recolonization are apparent even in Earth's early sedimentary rocks. The data have been controversial, with people affirming evidences for life and others rejecting it. The differences continue, but the increasingly-rigorous investigations are making the case for "life" robust. Dauphas et al. (2004) have looked at Early Archean banded rocks from Isua, Akilia, and Innersuartuut, Greenland. They find iron isotope enrichment diagnostic of microbial activity. There is no clear evidence that the Earth was ever barren of life. Wherever we look in sediments, we find traces that show living things actively colonizing—although, of course, in the Archean and Proterozoic, the environments were quite extreme.

These were conditions suitable for the development of stromatolites, formed by certain types of colonial bacteria and algae. Although stromatolites have been reported throughout the Precambrian, the tendency now is to identify abiogenic origins for some of these structures—particularly earlier in the Precambrian. This has been argued by Grotzinger and Rothman (1996) with commentary by Tyler (1997). It now seems fair to say that, without other evidence, stromatolites in all their 'diagnostic forms' can be generated as abiogenic crystal growth structures. Nevertheless, the case for in situ growth for *some* stromatolites is significant, and this is an argument for recognizing the passage of time in the Precambrian.

At this point it is necessary to comment on the preservation of marine organisms during and after the mabbul. Although the antediluvian land was destroyed, the ocean basins were affected only indirectly (large areas of the basins must have been made inhospitable to life). As a first approximation, these ocean basins may have represented 70% of Earth's surface area. Within the ocean basins were refugia: places where organisms were able to survive and live until conditions became suitable for them to migrate and recolonize the shallower waters of the flooded land surface.

Towards the end of the Proterozoic, there are signs of more familiar sedimentary structures occurring in the rocks, indicating that conditions were getting better for recolonization to take place. The first flowering of multicellular life forms comes with the Ediacaran fauna: soft-bodied life forms that have been the subject of much debate. In my opinion, these organisms were pioneers that could cope with conditions that would have killed other animals. They thrived for a time, but soon gave way to the main wave of recolonization.

Fossil embryos appear in the upper Proterozoic, as do surface-feeding organisms leaving horizontal burrows as trace fossils. Also found are the Vendian (at the top of the Proterozoic) small shelly fauna—indicative of continuing adverse environments. Nevertheless, taken together these fossils are pointers to conditions becoming more conducive to a wider range of life forms. They were preserved because the intense energies of the Hadean were diminishing. The Earth was then in a time of global catastrophism that could preserve the remains of fossils, if the organisms were available. These preserved organisms reveal the first stages of restoration of a food chain: bacteria, phytoplankton and zooplankton through most of the Precambrian, and multicellular extremophiles towards the end.

An Earth covered with water can be expected to develop unique patterns of erosion and deposition. Baumgardner and Barnette (1994) mathematically modeled oceanic water movements during a global flood with interesting results. Over continents, at high latitudes, strong cyclonic gyres appear with water velocities of 40–60 m/s. High-speed circular water flows would have eroded and sorted sediments, and formed a great diversity of deposits, ranging from meter-thick to kilometer-thick units. The recolonization approach allows sensitivity to the possibilities. For example, the Cambrian unconformity appears to be a horizon of widespread, but not global erosion (which may represent many local events that are correlated as one erosive horizon).

The Cambrian explosion marks a significant phase in the recolonization process. This was a time when there were large expanses of shallow shelf sea, with plenty of light and access to oxygen. Many animals could now invade these environments. Initial recolonization involved migration, and this took place at slightly different rates in different parts of the Earth. The controls were environmental rather than evolutionary. For example, Geyer (2005) documented body and trace fossils across the Proterozoic-Cambrian boundary in southern Namibia and showed the importance of considering local environments in understanding the faunal changes.

The speed of recolonization was facilitated by the relative lack of predators and by the availability of large expanses of uncolonized terrain. In addition, Brennan et al. (2004) report that seawater Ca^{2+} concentrations increased approximately threefold during the Early Cambrian, and that the timing of this change in seawater chemistry broadly coincides with the "Cambrian explosion." Organisms already possessing the capability to build exoskeletons would have taken full advantage of this environmental change. Leading up to the Cambrian there are many isotopic perturbations greatly exceeding any seen later (e.g., strontium, carbon, phosphorous) that had the potential to act as important environmental controls, facilitating sudden recolonization.

The abrupt appearance of so many phyla in low-energy environments (many Cambrian formations carry the marks of non-catastrophic deposition) makes little sense as an early (or the earliest) stage of the Flood. Neither does their sudden appearance make sense within Darwinian schemes because environmental changes alone cannot trigger the appearance of complex animals (Bengtson, 2004). Viewed on a recolonization canvas, the suddenness of the "explosion" was a direct consequence of widespread suitable habitats becoming available. The animals were able to move out of their refugia to colonize submerged continental areas as conditions ameliorated. The "explosion" restored the next section of the food chain, including molluscs, brachiopods, sponges, worms, jellyfish and arthropods (including trilobites). Contemporary with the faunal "explosion" was the "Cambrian substrate revolution," where barren sediments were transformed into habitats for benthic organisms (Bottjer et al., 1999; c.f., Seilacher and Pflüger, 1994)

The third stage of marine recolonization involved fishes. The first fishes were slower moving forms living close to the sea bottom, where they found their food. They were more easily overwhelmed and preserved by surges of sediment. Then, more and more species are recorded in later strata, arriving at the "age of fishes" in the Devonian. The variables influencing the fish fossil successions are yet to be elucidated, but may include seawater chemistry, the state of restoration of the food chain, and living environments. Predators followed and sharks became relatively common in the Carboniferous. Diversification within the created "kinds" led to speciation and eventually the development of modern forms. The keywords for all these changes (as monitored by the fossil record) is "ecology" and "variation within the basic type," not "evolution." Organisms appeared, multiplied and species diversified when conditions were right.

Terrestrial Recolonization

The book of Genesis informs us that, as a result of the mabbul, all air-breathing animals perished save those preserved on the Ark. Some plants must also have been carried along with the animals, but in most cases preservation was likely to be through floating mats of vegetation, seeds and spores. Recolonization of the land was initially by plants, with air-breathers appearing later as a result of dispersal from the Ark. Noah's experience with the olive leaf (Genesis 8:11) was, from the perspective of recolonization, not representative of the faunal regeneration of the Earth, but reflected a localized incident involving a plant able to grow for a time in a hostile environment.

It is then of particular significance when evidences of living, air-breathing animals are found in the rock record. The trackway data have been addressed by Garton (1991; 1996) and, in the context of the UK dinosaur trackways (Figures 1 and 2), by Tyler (2003b). The evidence to address is that the land is barren of air-breathers until the upper Paleozoic, although some amphibian trackways are known before this. Dinosaur tracks are widespread in the Mesozoic. Cenozoic and Quaternary strata bear witness to living birds and mammals.

Figure 1. Ardley Quarry dinosaur trackway. The tracks of numerous large dinosaurs are found in just one bedding plane, used as the floor of the quarry because of its hardness. The floor is covered by a thin lime mud, but after rain, the dampness in the tracks takes longer to dry out, facilitating visibility. The sediments are characteristic of shallow marine limestones.

Recolonization interprets this data in the context of recovering ecologies. The amphibian trackways belong to animals that are essentially water-dwellers (that could have survived outside the Ark). The dinosaurs are perceived as pioneers that were among the first to spread out over the unstable, emergent continental surfaces, soon followed by birds. The Tertiary trackways reveal recolonization by mammals in a world that looks increasingly modern in terms of its flora and fauna.

Plant recolonization was initiated by mosses and liverworts, followed by water-loving vascular plants, gymnosperms, ferns and then angiosperms and grasses. The damp and water-rich habitats represented by plants earlier in the succession are readily explained by the recolonization approach. The emergence of

Figure 2. Oxford Natural History Museum open-air exhibit of dinosaur trackway extracted from Ardley quarry.

angiosperms in the Cretaceous is likely to have been influenced by several factors associated with previously unsuitable growing conditions and the lack of birds to transport seeds. The late appearance of the grasses is surprising, although the recent discovery of five different species of grass in dinosaur dung (Prasad et al., 2005) indicates that we may not yet have an adequate database from which to develop interpretations. The grass fossil record, like the controversial Archean life record, is a reminder that 'absence of evidence is not evidence of absence'.

Since tracks made by living air-breathing animals (particularly dinosaurs) are a crucial point of difference between those diluvialists who interpret these traces as early-stage Flood deposits and those who interpret them as post-Flood recolonization, several publications have addressed these evidences directly or have considered rocks that were laid down contemporaneously with those of the trackmakers. Thus, for example, Garner (1996a) considered the evidence of dinosaur nests and also the Mesozoic flood basalt deposits (1996b). Scheven (1988) provided an overview of Cenozoic deposits and their interpretation in terms of post-Flood recolonization. Tyler (1996b) examined the remarkable development of chalk in the late Cretaceous and found a post-Flood interpretation to be far superior to alternatives.

Acceptance of Cenozoic strata as post-Flood is common to several Flood models. Nevins (1974) concluded that the John Day succession of Oregon, was post-Flood. In an accompanying commentary, Northrup (1974) observed: "Now if there is any one thing that I believe is certain about Cenozoic times, it is that they were post-Noahic." Brand (1997) recognized that patterns of Cenozoic sedimentation are quite different from those in the Paleozoic and Mesozoic, and specifically mentioned geological deposits that he interpreted as post-Flood. These included the Yellowstone fossil forests, the Wasatch and Bridger Formations, and the Green River Formation. In the case of the Yellowstone fossil forests (Figure 3), Brand drew heavily on the research of Coffin (1983; 1997). Brand (1997, p. 306) noted: "The upper Tertiary holds more direct evidence of bird and mammal activity that must have been postflood, including corkscrew-shaped burrows of extinct giant beavers." He drew attention to bird and mammal tracks and added: "Many of these footprints probably can be interpreted as postflood fossils" (pp. 306–307).

Figure 3. Polystrate tree stump from Specimen Ridge, Yellowstone National Park. The presence of roots has convinced many that this tree is in situ. However, a better description is that the tree has a root ball preserved. By analogy with deposits at Spirit Lake, Mt St Helens, such evidences point to transport prior to burial.

Field Geology

One of the challenges of historical geology is to interpret a diverse range of evidences to develop an integrated history of an area that satisfies all the data. Nicholaus Steno was the first geologist to recognize this and he attempted to develop principles that would make sense of the rocks of Tuscany in the context of biblical history. Today, diluvialists will make more of an impact when we show that the biblical

Figure 4. Basalt flow on Mull and McCulloch's tree. The polystrate tree is preserved as a 40-ft high cast with a small stump at beach level. Cooling has produced vertical columns in this basalt flow, but adjacent to the tree, the columns swing round and meet the tree at right angles. These observations are part of a story of catastrophism for the pile of flood basalts in north-west Scotland.

framework of history actually helps us to understand the rocks around us. The recolonization approach has a distinctive perspective on time, mingling the declining energy of global catastrophism with the drama of ecological successions.

Several papers have documented the value of recolonization for field studies. Tyler (1994) reviewed the local section in Jurassic rocks of the Yorkshire Coast and explored the potential of recolonization to explain the characteristics of the succession. Garton (1997) provided a field report of an organized excursion to the Rioja region of Spain, with an integrated story of Paleozoic, Mesozoic and Cenozoic sedimentation, and a consideration of the geological setting for the finest dinosaur trackways in Europe. Tyler (2003a) considered the basalts of the North Atlantic Igneous Province exposed on the Scottish island of Mull (Figure 4). Various geological features were reviewed to show the implications for time.

A striking example of post-Flood catastrophism comes from the Cerro Bianco region of Peru, where there is an unusual geological deposit called the Pisco Formation assigned to the Eocene. The upper part, about 80 m thick, has diatomaceous units plus mudstones with volcanic ash. Within this unit are fossil baleen whales, 5–13 m in length, many of which are extremely well preserved. In recently reported research by Brand et al. (2004), 346 whales were studied in a 1.5 km^2 area of exposed ground. These fossils occasionally preserved soft tissues, including baleen. The researchers noted the excellent preservation of many of these whales and inferred that rapid burial in diatomaceous sediments was essential to avoid the bones being disarticulated and dispersed (c.f., Tyler, 2004).

Fossil Successions

Many fossil successions show evidences of diversification: variation within morphological boundaries so that the end members are nevertheless recognizably related to the initial forms. Recolonization provides a coherent framework for understanding these fossil successions as variations within the basic type, more popularly referred to as variations within the created kind. There are real time sequences and real radiations but, of course, these changes are not dependent on Darwinian mechanisms of mutation and natural selection.

Micraster "evolution" within the chalk strata has been recognized by creationists to be variations within the basic type (Ward, 1971), but there is no rationale for any explanation of the trends within depositional models for chalk over a few weeks or even months.

Tyler's (1996b) post-flood scenario for chalk deposits provides a timescale that is long enough for successive generations of these animals to diversify according to environmental changes.

Garner (1999) considered the fossil horse data and concluded that the bones record a real diversification within the basic type Equidae subsequent to the Flood. Although the sequence is often used to support Darwinian evolution, the evidence is perfectly compatible with the creation model.

Robinson (2000b) looked at the fossil record of snakes and concluded that they can be considered as one basic type, with diversification (speciation) as a response to environmental change.

Conclusion

Like other diluvialist scenarios, the recolonization approach is committed to truth seeking (involving the proper handling of geological evidence) whilst upholding the witness of *both* Old and New Testaments. Like other diluvialists, Recolonizers have aimed to develop a biblically-based framework for developing geological science. The research of geologists (past and present) is approached positively, with the understanding that most professional people seek integrity and are genuinely interested in the accuracy of their field observations. The keyword for recolonization, however, is "ecology" not "evolution."

The approach provides some flexibility for assigning time to the fossiliferous sediments (Figure 5). Biblically, there is one geological period where we have clear chronological information: the Hadean, which lasted for up to 40 days. The time when Noah disembarked from the Ark is clear enough biblically, but there is no reason to think that this event can be correlated with one particular horizon in the rock record. All Recolonizers locate "leaving the Ark" earlier than the Permian, which is when the tracks of air breathing animals appear. As far

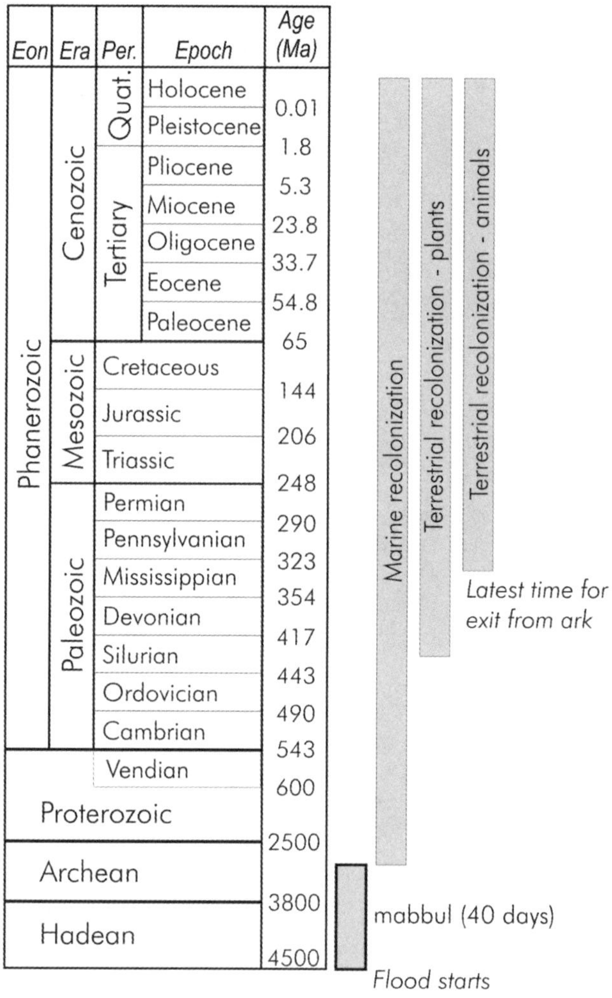

Figure 5. Timing of the Recolonization model relative to the geologic column. Timescale of the column provided for reference only.

as the rock record itself impacts on chronology, the geological evidences suggest short timescales: from minutes to decades. No observations of rock units require timescales measured in thousands of years or millennia.

The recolonization approach is still developing, with advocates exploring different ways of understanding the data. Nevertheless, it is an approach that is providing answers to the challenge of interpreting field observations. Recolonization finds special significance in the words of Jesus Christ about the

Flood: "As in the day of Noah so shall it be at the coming of the Son of Man." Our Lord speaks of the manner of His coming. In Noah's day, the waters did not rise gradually. The mabbul came unexpectedly and overwhelmingly. Along with the destruction of Sodom, Christ used these historical parallels because they were illustrations of how *unexpected*, how *rapid*, and how *comprehensive* will be the events in the hour when He returns. His witness to the Flood brings a unity to the character of God's 'end time' judgments in both Old and New Testaments.

References

CENTJ—*Creation Ex Nihilo Technical Journal*

CRSQ—*Creation Research Society Quarterly*

TJ—*Technical Journal (*formerly *CENTJ,* now *Journal of Creation)*

Austin, S.A. and K.P. Wise. 1994. The pre-Flood/Flood boundary: as defined in Grand Canyon, Arizona and eastern Mojave Desert, California. In Walsh, R.E. (editor). *Proceedings of the Third International Conference on Creationism*, pp. 37–47, Creation Science Fellowship, Inc., Pittsburgh, PA.

Baumgardner, J.R.. 1990. 3-D finite element simulation of the global tectonic changes accompanying Noah's Flood. In Walsh, R.E. and C.L. Brooks (editors). *Proceedings of the Second International Conference on Creationism*. Pp. 35–45. Creation Science Fellowship, Inc., Pittsburgh, PA.

Baumgardner, J.R. and D.W. Barnette. 1994. Patterns of ocean circulation over the continents during Noah's Flood. In Walsh, R.E. (editor). *Proceedings of the Third International Conference on Creationism*. Pp. 77–86. Creation Science Fellowship, Inc., Pittsburgh, PA.

Bengtson, S. 2004. The body-plan explosion, *Nature* 430:506 (29 July).

Bottjer, D.J., J.W. Hagadorn, and S.Q. Dornbos. 2000. The Cambrian substrate revolution. *GSA Today* 10(9):1–7.

Brennan, S.T., T.K. Lowenstein, and J. Horita. 2004. Seawater chemistry and the advent of biocalcification. *Geology* 32(6):473–476.

Brand, L. 1997. *Faith, reason and Earth history*. Andrews University Press, Berrien Springs, MI.

Brand, L.R, R. Esperante, A.V. Chadwick, O.P. Porras, and M. Alomía, M. 2004. Fossil whale preservation implies high diatom accumulation rate in the Miocene–Pliocene Pisco Formation of Peru. *Geology* 32(2):165–168.

Cleland, C.E. 2001. Historical science, experimental science and the scientific method. *Geology* 29(11):987–990.

Coffin, H.G. 1983. *Origin by Design*. Review and Herald Publishing Association, Washington.

———. 1997. TheYellowstone petrified "forests." *Origins (Journal of the Geoscience Research Institute)* 24:5–44.

Dauphas, N., M. van Zuilen, M. Wadhwa, A.M. Davis, B. Marty, and P.E. Janney. 2004. Clues from Fe isotope variations on the origin of early Archean BIFs from Greenland. *Science* 306:2077–2080.

Dutkiewicz, A., B. Rasmussen, and R. Buick. 1998. Oil preserved in fluid inclusions in Archaean sandstones. *Nature* 395:885–887.

Garner, P. 1996a. Where is the Flood/post-Flood boundary? Implications of dinosaur nests in the Mesozoic. *Creation Ex Nihilo Technical Journal* 10(1):101–106.

Garner, P. 1996b. Continental flood basalts indicate a pre-Mesozoic Flood/post-Flood boundary. *Creation Ex Nihilo Technical Journal,* 10(1):114–127

Garner, P. 1999. It's a horse, of course. *Origins (Journal of the Biblical Creation Society)*, No. 26, October, 13–23.

Garton, M. 1991. Rocks and scripture: on the right track? *Origins* 4(10):3–8.

———. 1996. The pattern of fossil tracks in the geological record. *CENTJ* 10(1):82–100.

———. 1997. A Spanish weekend. *Origins* 22:11–24.

Geyer, G. 2005. The Fish River Subgroup in Namibia: stratigraphy, depositional environments and the Proterozoic–Cambrian boundary problem revisited. *Geological Magazine* 142(5):465–498.

Grotzinger, J.P. and D.H. Rothman. 1996. An abiotic model for stromatolite morphogenesis, *Nature* 383:423–425.

Harrison, T.M., J. Blichert-Toft, M. Müller, F. Albarede, P. Holden, and S.J. Mojzsis. 2005. Heterogeneous Hadean hafnium: evidence of continental crust at 4.4 to 4.5 Ga. *Science* 310:1947–1950.

Hunter, M.J. 1996. Is the pre-Flood/Flood boundary in the earth's mantle? *CENTJ* 10(3):344–357.

Kamber, B.S., M.J. Whitehouse, R. Bolhar, and S. Moorbath. 2005. Volcanic resurfacing and the early terrestrial crust: Zircon U–Pb and REE constraints from the Isua Greenstone Belt, southern West Greenland. *Earth and Planetary Science Letters* 240(2):276–290.

Knauth, L.P. 2005. Temperature and salinity history of the Precambrian ocean: implications for the course of microbial evolution. *Palaeogeography, Palaeoclimatology, Palaeoecology* 219(1–2):53–69.

Knauth, L.P. and D.R. Lowe. 2003. High Archean climatic

temperature inferred from oxygen isotope geochemistry of cherts in the 3.5 Ga Swaziland Supergroup, South Africa. *Geological Society of America Bulletin* 115(5):566–580.

Melezhik, V.A., A.E. Fallick, E.J. Hanski, L.R. Kump, A. Lepland, A.R. Prave, and H. Strauss. 2005. Emergence of an aerobic biosphere during the Archean-Proterozoic transition: challenges of future research. *GSA Today* 15(11):4–11.

Nevins, S.E. 1974. Post-Flood strata of the John Day country, northeastern Oregon. *CRSQ* 10(4):191–204.

Northrup, B.E. 1974. Comments on the Stuart E. Nevins paper (Post-Flood strata of the John Day country, northeastern Oregon). *CRSQ* 10(4): 205–207, 228.

Ohmoto, H. 1996. Evidence in pre-2.2 Ga paleosols for the early evolution of atmospheric oxygen and terrestrial biota. *Geology* 24(12):1135–1138.

Prasad, V., C.A.E. Strömberg, H. Alimohammadian, and A. Sahni. 2005. Dinosaur coprolites and the early evolution of grasses and grazers. *Science* 310:1177–1180.

Rosing, M.T. and R. Frei. 2004. U-rich Archaean sea-floor sediments from Greenland–indications of >3700 Ma oxygenic photosynthesis. *Earth and Planetary Science Letters* 217(3–4):237–244.

Rosing, M.T., D.K. Bird, N.H. Sleep, W. Glassley, and F. Albarede. 2006. The rise of continents—an essay on the geologic consequences of photosynthesis. *Palaeogeography, Palaeoclimatology, Palaeoecology*, 232(2–4):99–113

Robinson, S.J. 1996. Can Flood geology explain the fossil record? *CENTJ* 10(1):32–69.

———. 1998. The Flood in Genesis: what does the text tell geologists? In Walsh, R.E. (editor). *Proceedings of the Fourth International Conference on Creationism*. pp. 465–474. Creation Science Fellowship, Inc., Pittsburgh, PA.

———. 2000a. The then world with water having been deluged perished. *Origins* 29:15–24.

———. 2000b. Snake evolution and the Garden of Eden. *Origins* 30:10–13.

Scheven, J. 1988. *Mega-sukzessionen und klimax im Tertiaer: katastrophen zwischen sintflut und eiszeit*. Haenssler-Verlag, Neuhausen-Stuttgart.

Seilacher, A. and F. Pflüger. 1994. From biomats to benthic agriculture: a biohistoric revolution. In Krumbein, W.E. D.M. Paterson, and L.J. Stal. (editors). *Biostabilization of sediments*, pp. 97–105. Bibliotheks und Informationssystem der Carl von Ossietzky Universität Oldenberg, Oldenberg.

Tyler, D.J. 1994. Tectonic controls on sedimentation in rocks from the Jurassic Series (Yorkshire, England). In Walsh, RE. (editor). *Proceedings of the Third International Conference on Creationism*. pp. 535–545. Creation Science Fellowship, Pittsburgh, PA.

Tyler, D.J. 1996a. Abiotic origin of stromatolites. http://www.biblicalcreation.org.uk/scientific_issues/bcs075.html (Accessed January 2006).

———. 1996b. A post-Flood solution to the chalk problem. *CENTJ* 10(1):107–13

———. 1997. The atmosphere of the Early Earth.http://www.biblicalcreation.org.uk/scientific_issues/bcs074.html (Accessed January 2006).

———. 2002. The "smoking gun" way to do geological science. *Origins* 35:12.

———. 2003a. Mull's witness to the Flood and its aftereffects. *Origins* 33:16–18.

———. 2003b. Tracking dinosaurs in Britain. *Origins* 36:6–9.

———. 2004. A whale of a tale. *Origins* 39:11.

Walker, T. 2004. Peperite: more evidence of large-scale watery catastrophe. *TJ* 18(1):18–21.

Ward, R.R. 1971. A study of English *Micraster* research. In Lammerts, W.E. (editor). *Scientific Studies in Special Creation*. pp. 184–197. Presbyterian and Reformed Publishing Company, Philadelphia, PA.

Whitcomb, J.C. and H.M. Morris. 1961. *The Genesis flood*. Presbyterian and Reformed Publishing Company, Philadelphia, PA.

Wise, K.P. 2002. *Faith, form and time*. Broadman and Holman Publishers, Nashville, TN.

Author

David J. Tyler has a background in physics (BSc and MSc) and a career that has led him into management (PhD). He has 8 years experience in applied research plus other industrial experience. As an academic, he is involved in teaching and research. His first paper in the *CRSQ* was in 1979 and he has been actively writing and speaking since then. He coordinates a group in the UK that is actively seeking to develop geological science built on biblical foundations. He is the author of *Creation—Chance or Design?"*

Forum Discussion

1. How can you support your interpretation of Scripture that all land dwellers were totally annihilated with no fossil trace on the first day of the Flood? Analogies to Sodom and the Second Coming seem unconvincing. (Oard)

The testimony of Scripture is actually very strong. Not only were all air breathers to perish in the Mabbul, but also the land was to be destroyed. This was a judgment upon the entire domain over which Adam had been given dominion. It was a destruction unparalleled in the history of the Earth. It is a fitting precursor to the Second Coming, when everything will be burnt up and there will be a new heaven and a new earth. (The "no fossil trace" on the first day of the Flood is not a position I advocate or defend).

2. Your chapter affirms uniformitarian interpretation, except for the timescale; e.g., the first paragraph of "Concluding Comments." Don't you think that catastrophism renders many of those interpretations invalid, demanding alternatives? (Oard)

This chapter does not affirm "uniformitarian interpretation." Uniformitarianism should be opposed as an unwarranted theoretical imposition upon the historical sciences. There is a similar danger of imposing catastrophism as a theoretical dogma. Our need, as biblically-based scientists, is to test ideas, explore alternatives that are informed by revelation, and be prepared to follow all the evidence wherever it leads.

3. If the Flood/post-Flood boundary lies in the middle Paleozoic, how can all the overlying strata with its "fossil successions" be deposited by post-Flood catastrophism? (Oard)

First, I welcome the reference to "fossil successions." A Flood model that seeks to address the evidence must do justice to the real patterns in the fossil record. I think the problem perceived in this question arises from the widespread view that the base of the Flood lies close to the Great Unconformity at the base of the Paleozoic. The Recolonization approach presented in this chapter affirms that the Flood deposits include all the Precambrian rocks. These are very extensive. The implication is that the Flood year has produced the bulk of the sedimentary and igneous rock record.

4. Contrary to claims, there is no unambiguous biblical support for destruction in one-day, total crustal overturn, or a 40-day "mabbul." The chronology of the Flood in Genesis affirms universal death after a minimum of 40 days of inundation (Genesis 7:21–23). (Reed, Klevberg, Froede)

The Recolonization approach is not tied to "destruction in one day," but it does give weight to the necessity of the Ark to be sealed by the hand of God on that first day for its passengers to be safe. Justice has to be done to the scriptural teaching about the destruction of the land, and no one can accuse the Recolonization approach of failing to address this. The 40-day Mabbul is very important for understanding the Flood narrative and its theological implications. There can be little doubt that the death of air-breathers was complete before the end of the 40-day inundation.

5. If the Flood's destruction was limited to the Hadean, why is there so much evidence of "destruction" (catastrophic fossilization, mountain building, ocean basin forming, sedimentation, etc.) later in the rock record and thus distinct from the Flood? (Reed, Klevberg, Froede)

To say that the Flood's destruction was "limited to the Hadean" is a gross understatement. The Hadean bears witness to an unparalleled destruction involving crustal reworking, and the limited evidence we have is that all this took place under water. There is every reason to think that extensive geological activity (of a lower energy) would have followed this global destruction. Those who advocate Recolonization do not think that this catastrophism as "distinct from the Flood" but as triggered by it.

6. If the entire crust was overturned in one day, would not heat flow (exposed temperatures of 1200°–2400° C) have destroyed life (including those in the ark), the ocean, and the atmosphere? (Reed, Klevberg, Froede)

The Hadean lasted up to 40 days, not "one day." Yes, heat flows would have been extreme. Yes, living things would have been destroyed (which is why no one need struggle with the Recolonization explanation for the absence of fossils derived from the pre-Flood land). Yes, the effects of this would have extended out to sea. The only place of refuge for air-breathers was the Ark. Since the sea was not destroyed, we are justified in postulating marine refugia.

7. The timing of the Flood and subsequent history places severe restraints on the migration, reproduction, and fossilization of life preserved in the "refugia" and the ark. (Reed, Klevberg, Froede)

Yes, migration takes time, but this is recognized within the Recolonization approach. Reproduction takes time and that is also recognized, being the key to understanding many of the fossil successions. Fossilization occurred at intervals during the recolonization process and was rapid, not providing an additional constraint on time. Marine recolonization followed a radically different course from recolonization by air-breathers. Recolonization applies the well-established biological principle that any disturbed ecosystem goes through phases of recovery. All this can take place within time frames long-accepted by the creationist community (e.g. Appendix 7 of "The Genesis Flood").

six

A Long and Winding Way

Emil Silvestru

Introduction

Geology has in many ways pioneered the use of extensive visual representations to convey compacted information, especially the relationship between time, rocks and fossils. The most comprehensive such graphic representation is definitely the geologic timescale[1] (GTS) which presents an idealized order in which evolutionary biology claims life has evolved. I am convinced that, although it is labeled "geologic timescale" this grand icon of geology is in fact nothing but a "subsidiary" of the grand icon of evolution: the "tree of life." By definition then, whatever the GTS presents, it does that with evolution in mind (i.e., an external bias) and therefore its objectivity is questionable from the very beginning.

The GTS incorporates a tremendous amount of primary and interpreted scientific information, most of which is accessible only to the initiated. To the average person it is the orderly and precisely placed divisions and subdivisions of time, all bearing unusual names, which seems to convey profound knowledge and certainty of interpretation. However, at a closer inspection one notices the graphic representation is counter-intuitive (Figure 1). The time factor (effective duration) is strongly skewed towards the shorter time units, (it is usually represented as a logarithmic scale) divided into more abundant subunits which need more graphic space to be represented. The actual time duration is expressed in numbers whose significance is missed by most users. Since the visual part is the most effective in conveying the intended information, the vast majority of people will be left with a wrong perspective of the time dimension involved (Figure 2). But that perspective is deeply

[1] GTS is used synonymously with "geologic column" or "geological column" in other chapters.

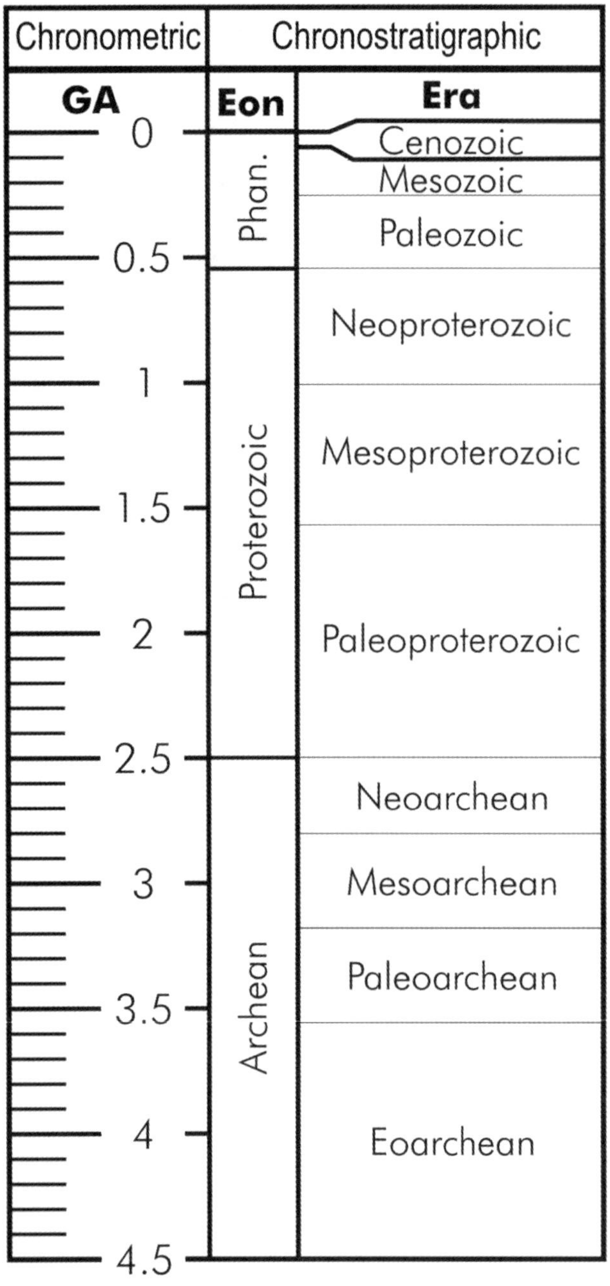

Figure 1. The GTS as it would appear scaled correctly to time.

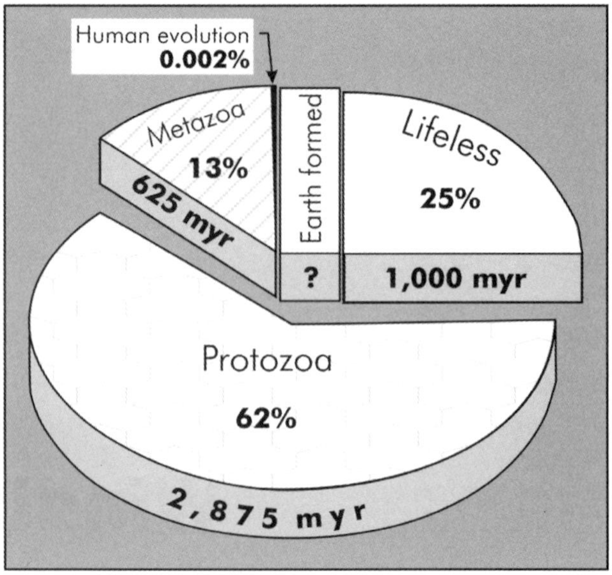

Figure 2. A pie chart showing the volume of time in each eon of the GTS.

imprinted in the public memory and I believe it is not an exaggeration to say that the GTS is one of the most powerful secular icons of today.

Because of all this, the attempt to redefine or even replace the GTS represents a very serious challenge to creationary geology (I will be using the adjective "creationary" as the antonym for "evolutionary"), especially since unlike the iconoclasts in the 8th century, smashing the icon will not solve the problem; it will have to be replaced with another.

Since I am the only one in this exquisite outfit of creationists that has been educated and "formatted" in Central-Eastern Europe I will deliberately avoid analyzing in detail any existing creationary model and will not suggest any alternative one. Rather, I would like to provide an external, "third part" view which may be useful for this endeavor.

What is Creation Science All About?

I believe that there is need to properly answer this question first, in order to understand the real challenges in demolishing a well-established evolutionary icon. There seems to be a relaxed attitude among creationary scientists as to what their scientific approach is all about. Some seem to regularly refer to the Bible as source of primary information; others tend to focus on non-biblical data. Historical information from the Bible cannot and should not be classified as "theological" or "supernatural" (as opposed to "natural"). Reed and Froede use the expression "creationist natural history" (2003, p. 91) as opposed to "naturalist" natural history. This however raises a problem: for the present evolutionary establishment "natural history" does not necessarily mean "history of nature" but also reflects their commitment to interpreting the history of the universe in terms of natural causes. Hence the main tenet of evolutionary philosophy: science can only mean explaining things by natural causes. Understanding the futility of a public debate on this particular definition, the pioneers of modern creationism have tried to reinterpret the facts of nature by way of the same rules (logic, natural laws, experiments and predictions) so that they will agree with the young-earth framework the Bible describes. In this sense, "creationist natural history" is a correct description. However, the new generation of creationary scientists has taken investigation further, trying for example to scientifically explain Creation itself. That means explaining the supernatural, because humans can only understand the universe by way of the natural laws, but those were not in place at the moment of Creation; they were only established during it. Once the natural laws were in place, it became possible for humans to scientifically understand Creation. Geology essentially deals with the post-Creation period and one can therefore rightfully talk about a "creationary geology" as scientific knowledge.

Creationary Geology and the Geologic Timescale

To some creationary geologists, proving the reality of the Genesis Flood has mainly to do with proving the accuracy of the biblical narrative. To me—since my main goal is to explore the avenues of geology—it has to do mostly with exploring another possible geological setting with all its implications. I believe that in this context, proving the accuracy of the Bible should be more a conclusion, than a purpose per se. If so, the question arises: all biblical information aside, what are the paradigmatic alternatives of interpreting the rock and fossil records? In terms of time/duration, the two records can be interpreted as the result of:

1. Slow, uniform processes over a long time.
2. A global catastrophe (flood essentially) over a short period of time.
3. Slow, uniform processes over long time, with spikes of regional/local catastrophic events.

In the evolutionary establishment there is a clear shift from "(1)," the Lyellian position/dogma to "(3)," neo-catastrophism. This position though, comes with yet another one of those contradictory attitudes so typical to the establishment: although the GTS is the offspring of the Lyellian dogma, it is preserved virtually unchanged within the neo-catastrophic paradigm. However, this inconsistency does not seem to bother most geologists, paleontologists, and biologists. Some however, acknowledge the problems and the need for change:

> However, there is now an appreciation of the importance of both normal and catastrophic sedimentation, and of the *necessity to distinguish between them* in the rock record. The distinction may not always be easy, *but must be attempted if correct interpretations of the processes of sedimentation and the controlling factors are to be made* (Reading, 2002, p. 17, emphasis mine).

There are some deeper implications in this quote which will be mentioned later. Young-earth creationists on the other hand, who believe that the second paradigmatic alternative is the true interpretation of the rock and fossil records, cannot properly exploit this inconsistency because of the significant disagreement regarding the position of the catastrophe in the GTS (which represents, for the time being, the only complete geological list of known rocks and fossils). Virtually all creationary geologists acknowledge that the biblical Flood, which should be seen in the context of this discussion as a geological not a theological event, was the catastrophic event that caused most of the sedimentary rock record (the source for the GTS). But how much of it, or in other words, where the pre- and post-Flood boundaries are to be found in the GTS, is the cause of the disagreement.

Some of the problems can be traced back to a foundational difference between the young-earth and evolutionary views of the fossil and rock record, namely the type of accuracy involved in the division of geologic time. The former view is based on one major event covering roughly a year thus needing an "archeological degree of accuracy," the latter is based on many (cyclical or not) events covering billions of years, hence a "geological degree of accuracy" (Silvestru, 2004, p. 216). While there is general agreement amongst creationary geologists that the assumptions on which the GTS is based are faulty, there is a great deal of debate on whether to reinterpret the GTS or completely replace it. Some creationary geologists believe that the GTS should be reinterpreted within a young-earth paradigm. Some believe that the only solution is total rejection and the replacement with an alternative, Flood-based table.

The complete replacement proposed by Reed and Froede (2003) for example, would be fine if the only purpose of creationary geology is to create an alternative geology, completely disregarding the GTS and evolutionary geology. The trouble I see with such an approach is that at some point the correlation between the two models will be inevitably requested by our adversaries and supporters alike. The former would be justified in asking where the pre- and post-Flood boundaries are in the GTS because what they would understand is that a long-established model would be completely disregarded without any proper explanation and critique (as far as they are concerned) and that is inadmissible in science. Furthermore, that would allow secular geologists to once again accuse us of lack of scientific criteria. The latter would request the correlation because they will need to be able to counter their critics whenever accused of accepting an unscientific model. It is infinitely more difficult to defend one's position by proposing a completely new model than it is to reinterpret an existing model. Consequently, whether an agreement on the boundary issue is reached before or after the hypothetical alternative to the GTS is brought forth, such an agreement must be reached. If so, it makes more sense (to me at least) to reach this agreement before an alternative to the GTS is attempted.

Another problem which needs to be dealt with is how much of the existing evolutionary data will be used in order to assemble an alternative timescale. I cannot see how creationary geologists would be able to revisit all significant locations and occurrences on the planet, in order to come up with their own interpretation of the field data. On the other hand, I do not believe that using the geology of a single continent (North America whence most of the alternative, creationary field studies come from) is acceptable. Global data needs to be integrated into an alternative to the evolutionary GTS. So, it seems that evolutionary data will have to be accepted without being verified in the field. What would the criteria be? Is there an agreement on this? Again I believe such an agreement will have to be reached before an alternative model is proposed. There are numerous examples within the evolutionary interpretation of the rock and fossil records in which diagnosis is questioned:

1. The higher grade metamorphites lying on top of lower grade ones. Some interpret such a superposition as anomalous (postulating that metamorphic grades increase with depth as a general rule) and have to invoke overthrusts. I have spent seven years in the field as prospecting geologist in northern Romania and have investi-

gated such "anomalous" settings through mining and drilling. Not a single time the geometry and predictions consistent with nappe/overthrust tectonics were confirmed. Retromorphism (or retrograde metamorphism) on the other hand (the chemical down-grading of a homogenous metamorphic sequence) explained concrete settings very well. One cannot stop wondering how deep such chemical changes can actually be involved in the very fabric of metamorphic domains.

2. The interpretation of various diamicts as either tillites or submarine landslides (Oard, 2000).
3. Oil & gas being a fossil or a primary resource. The latter was a view held by Mendeleyev (the father of the periodic table of elements) and in recent times by Thomas Gold (1999). Although it is extremely controversial today, the idea that hydrocarbons are in fact primary constituents of the planet is gaining more and more support as new data accumulates.
4. Paleokarst in the form of sinkholes (subaerial) which can also be interpreted as ancient submarine pockmarks (Silvestru, 2001).

Rather than continuing with other examples, I believe a quote on the concept of "facies" from a recent textbook would better summarize this point:

> For example we can attempt to define objectively a sedimentary product, for example 'cross-bedded ooid grainstones,' 'red, parallel-bedded sandstone facies;' or we can subjectively interpret a process, for example, 'turbidite facies,' meaning the we *believe* it to have been deposited by turbidity currents, not that we can ever be sure it *was* deposited by turbidity currents. If we use an environmental term such as 'basinal facies,' 'platform facies,' 'fluvial facies,' it means that it has the characteristics compatible with such an environmental interpretation, but clearly, nevertheless, it is only an interpretation (Reading, 2002, p. 19, emphasis his).

Reading's point raises interesting questions for creationists about lithostratigraphy and/or biostratigraphy. Should we discard both? Can we preserve the "objective" (in the quote) part of the lithostratigraphic column? I believe we can if it would only be a sequence of basic types of rocks with no initial sedimentary environment connotations. Such a column (preserving at least a part of the standard names) would be usable even if most of it were deposited by lateral movement (Berthault, 1994) rather than by vertical settling (the vertical sequence is still a function of the lateral sequence only that is not a time function but a flow dynamics one). The environmental significance would eventually be discerned as a result of dynamics correlations. This may raise the problem of correlations within the GTS (which are also based predominantly on fossil contents). Yet this should not be too much of a problem since there are many convergent sedimentary sequences with regional or even global extent that can be used as markers. For example the interval between the Late Permian and the Early Jurassic in Europe is characterized by the repetition of a sequence of reddish conglomerates, sandstones, red siltites, limestones, twice. In places, the similarities between the Early Triassic (Scythian) siltites and the Early Jurassic (Hettangian-Sinemurian) ones are so strong and the lack of fossils so wide-spread that the only way to diagnose the formations is mineralogy and the larger stratigraphic context. In fact at many locations on Earth, this interval in the GTS is characterized by more or less similar formations. Such features should provide good correlative markers while the formations between them could easily be redefined.

Though the absolute majority of stratigraphic divisions in the GTS are delimited on the grounds of fossil contents, there are several exceptions:[2]

1. The base of the Ediacaran system: termination of the Varanger (or Marinoan) glaciation.
2. Lower Cambrian series: near base of negative carbon isotope excursion.
3. Base of Aptian stage: magnetic polarity chronozone, base of M0r.
4. Base of Danian stage: iridium geochemical anomaly.
5. Base of Thanetian stage: magnetic polarity chronozone, base of C26n (a temporary assignment).
6. Base of the Ypresian stage: base of negative carbon-isotope excursion.
7. Base of the Aquitanian stage: base of magnetic polarity chronozone C6Cn.2n.
8. Base of the Messinian stage: astrochronology age of 7.251 Ma; middle of magnetic polarity chronozone C3Br.
9. Base of Pliocene series: top of magnetic polarity chronozone C3r, ~100 kyr before Thvera normal-polarity subchronozone (C3n.4n).
10. Base of Piacenzian stage: base of magnetic polarity chronozone C2An (Gauss).
11. Base of the Gelasian stage: isotopic stage 103, base of magnetic polarity chronozone C2r (Matuyama).
12. Base of middle Pleistocene series: Brunhes-Matuyama magnetic reversal.
13. Base of upper Pleistocene series: base of the Eemian interglacial stage (base of marine isotope stage 5e) before final glacial episode of Pleistocene.

It may be argued that even these non-biological markers are questionable, but the fact that physical characteristics of the rocks can be used to define boundaries remains. The lunar geologic timescale for example, (all long age speculations aside) was built on the density of impact craters and volcanic activity (Dalrymple, 2000) with no significant sedimentary assumptions and obviously no fossils.

The fact that there are alternative non-evolutionary interpretations of certain rock formations within the GTS is useful for a diluvial approach of the topic. Yet, there will be many situations in the attempt to create a young-earth GTS in which alternatives will not be available in the literature and building them on literature only (without direct observation) may not be accurate enough. Discarding them altogether is not acceptable either, because all existing rocks should be eventually accountable for in an alternative model. Consequently, the use of data from the standard GTS seems to me unavoidable. What would then the criteria be for accepting it? Some have suggested accepting the GTS but concentrate on the rate of accumulation of the sediments and fossils (Snelling, 1996). Others (as mentioned before) have suggested rejecting the GTS completely and building an alternative based on "geologic energy" of sedimentary processes during the Flood (Reed et al., 1996).

Of course such positions raise another issue on which the creationary geology community is still divided. What was the main geological mechanism that caused the Flood? There are significant differences in the dynamics (and energy patterns for that matter) according to the geological and geophysical model one uses for the Flood. And the resulting rocks and fossils would normally have to fit the respective dynamics.

Using geologic energy as a criterion for assessing the rock record—a position I subscribe to—comes in fact very close to what Reading said (see previous quote) about discerning between normal and catastrophic sedimentation. And since there seems to be an interest in the evolutionary establishment (at least amongst some sedimentologists) in following

[2] Cited from the International Commission on Stratigraphy.

this avenue, I believe we should keep a close eye on the research published (though identifying the useful information from the evolutionary background noise could be a difficult task!). It may well help our efforts.

A Model or a Column?

One necessary clarification I see is the strategic approach: should we first replace the GTS with a diluvial alternative and then, based on it, build a geological model of the Flood? Or should we first try to agree on a general framework model/paradigm of the Flood and then interpret/replace the GTS based on that model?

From a lithological point of view, the GTS was mostly the result of the latter approach since all rocks were interpreted as the result of processes still operating today, or the existing model of uniformitarianism. In our case, we are postulating a unique geologic event which has generated the vast majority of rocks, and some of those rocks may not be forming today. Hence, I believe we need to first create a framework model and then proceed with the reinterpretation/replacement of the GTS.

A Possible Strategy

Unlike evolutionary geology (which is based on a set of universally accepted assumptions in a logical sequence even if **those assumptions are wrong in my view**), creationary geology's only universally accepted assumption is that the Flood was a real geologic event. I would like to emphasize again that in my view, a professional approach without recurring to the Bible as a source of primary geological information is perfectly possible and it should be followed first. Otherwise our endeavor would be labeled "biblical creationism" as opposed to "scientific creationism." Therefore, if we cannot come to at least a minimum set of accepted assumptions in a logical and hierarchic sequence, trying to replace the GTS will be nothing but a loss of time, resources and energy. I would see a possible set of such assumptions as follows:

1. The Flood was the result of *"the following"* (to be determined) process(es).
2. It unfolded in *"the following"* sequence of geologic events.
3. The pre-Flood boundary is marked by *"this particular"* global feature.
4. The Flood/post-Flood boundary is marked by *"this particular"* global feature.

These global features may not be recognized in the GTS today as such, but they should be recognizable and capable of being correlated globally. In my opinion, since the Flood was global both its onset and final stages, it should have produced globally recognizable features like unconformities. These features therefore should be the very first objective of our efforts. Though fossil contents should by no means be the main criteria in identifying them, fossils could still be used for correlation once evolutionary assumptions were discarded. The road ahead of us is long and winding, so the sooner we start walking on it together, the better.

Having been involved in the past in somewhat similar efforts within the karstological community, I believe that the best starting point would be an informal meeting of all those interested with the sole purpose of identifying and agreeing on such basic features and issues. That would significantly shorten the inception phase.

References

CRSQ—*Creation Research Society Quarterly*

Berthault, G. 1994. Experiments in stratification. In Walsh, R.E. (editor), *Proceedings of the Third International Conference on Creationism* (technical symposium sessions), pp. 103–110. Creation Science Fellowship, Pittsburgh, PA.

Dalrymple, G.B. 2000. *The Age of the Earth*. Stanford University Press, Stanford, CA: 219.

Gold, T. 1999. *The Deep Hot Biosphere.* Springer-Verlag, New York.

International Commission on Stratigraphy at: http://www.stratigraphy.org/ (June 2004).

Oard, M.J. 2000. *Ancient Ice Ages or Gigantic Submarine Landslides.* Creation Research Society Books, Chino Valley, AZ.

Reading, H.G. (editor). 2002. *Sedimentary Environments: Processes, Facies and Stratigraphy,* third edition with corrections. Blackwell Science, Malden, MA.

Reed, J.K., C.R. Froede, Jr., and C.B. Bennett. 1996. A biblical Christian framework for earth history research: part IV—the role of geologic energy in interpreting the stratigraphic record. *CRSQ* 33(2):97–101.

Reed, J.K., and C.R. Froede, Jr. 2003. The uniformitarian stratigraphic column–shortcut or pitfall for creation geology? *CRSQ* 40:90–98.

Silvestru, E. 2001. The mystery of paleokarst solved. *Creation Ex Nihilo Technical Journal* 15(3):112 (now *Journal of Creation*).

———. 2004. Descending into the pit to avoid the fall (response to Reed and Froede). *CRSQ* 40:216–219.

Snelling, A.A. 1996. Where should we place the Flood/post-Flood Boundary in the Geological Record? *Creation Ex Nihilo Technical Journal* 10(1):30 (now *Journal of Creation*).

Author

Emil Silvestru graduated in geology at the "Babes-Bolyai" University in Cluj, Transylvania, Romania in 1979, where he also earned his MSc and PhD studying the sedimentology of karst terrains before becoming an Associate Professor. For seven years he prospected (mining and drilling) for polysulfide ore deposits in metamorphic formations. He specialized in karstology and worked for 12 years at the world's first speleological institute in the city of Cluj; eight years as its director. He has published 39 research papers and co-authored a book. Since 2002 he has worked as a senior scientist and speaker for Creation Ministries International in Canada (formerly Answers in Genesis).

Forum Discussion

1. In "Creationary Geology and the Geologic Timescale" (paragraph 5) you state that if any alternative interpretation of the rocks will have to correlate with the geological column. Would not the correlation between Flood events and the geological column be variable—some rocks (e.g., banded-iron formations) may correlate with the beginning of the Flood, but other rocks, such as the Cenozoic, may not correlate with the end of the Flood. Why should we be concerned with such a correlation and any demands of evolutionary, uniformitarian scientists in this regard (such as we need to be "scientific")? (Oard)

If we will be using a completely different model, people would inevitably ask things like: "where does the Late-Jurassic bauxite-bearing alleged paleokarst fit in your model?" or "where does the Permian extinction fit in your model?" etc. Also, wherever we will end up placing the Flood boundaries, they will also have to be expressed in terms of the GTS, will they not?

2. Even if it requires much work and time to propose an alternative model, should we not propose one anyway, highlighting the mechanism (the Flood) that laid down the vast majority of the rocks and fossils? (Oard)

I do not believe that a replacement of the GTS would be acceptable without mentioning the boundaries in it and just saying that wherever those boundaries would eventually be, we emphasize the mechanism rather than the its final product–the sequencing of rocks and fossils

3. In "A Possible Strategy" you suggest that it might be possible to construct a Flood model without reference to the Bible. Why should we, and why can't we use a geological framework based on scripture, such as Walker's model? (Oard)

Because I still believe we are trying to reinterpret geology based on the same facts and field data. Once we use the Scripture as the frame source of information we are more biblical creationists rather than scientific creationists. My understanding is that we mean to replace a geologic model with another.

4. Drawing a distinction between the rock record and the geologic column (Chapter 8) would ease the author's tension between axiomatic flaws in the GTS and his desire to pursue stratigraphy. (Reed, Klevberg, Froede)

That may be, but for me the main drive to do something does not come from tension between axiomatic flaws. It rather has to do with my need to interpret what I see and what I learned before I was a creationist with the minimum of changes and the maximum of results so that I would feel satisfied as a geologist and as a person. If that is not possible (though I think it is) I may consider a completely new paradigm.

5. The use of non-evolutionary local and regional "markers" is helpful, but should not be confused with using the column, and must be empirically tested for stratigraphic uniqueness and lateral extent. (Reed, Klevberg, Froede)

My point was that there are examples of non-biostratigraphic markers even in the GTS: "but the fact that physical characteristics of the rocks can be used to define boundaries remains." If we will use those markers or others, the fact that one can attempt building a timescale without any evolutionary assumptions remains.

6. Would the author's hypothesized global unconformity cross the time lines of the geologic timescale? If so, what does that say about the timescale? (Reed, Klevberg, Froede)

I find it difficult to believe that such a major geological feature would have went unnoticed by geologists over two centuries of field work. It is its real meaning that was missed though. Just like with the famous iridium-rich K/T boundary layer: is it due to volcanism or impact? Therefore, if it is already part of the GTS but for all the wrong reasons, all it would say about it is what we all know: the GTS contains a lot of wrongly interpreted real field data.

7. You indicate that a column based on lithostratigraphy may be viable and that sequences with regional or global extent might be used as markers. Could you point to some specific cases where such correlations are possible? (Tyler)

I though I did that when I wrote: "Yet this should not be too much of a problem since there are many convergent sedimentary sequences with regional or even global extent that can be used as markers. For example the interval between the Late Permian and the Early Jurassic in Europe is characterized by the repetition of a sequence of reddish conglomerates, sandstones, red siltites, limestones, twice. In places, the similarities between the Early Triassic (Scythian) siltites and the Early Jurassic (Hettangian-Sinemurian) ones are so strong and the lack of fossils so widespread that the only way to diagnose the formations is mineralogy and the larger stratigraphic context. In fact on many locations on earth, this interval in the GTS is characterized by more or less similar formations. Such features should provide good correlative markers while the formations between them could easily be redefined.

8. If a lithostratigraphic column is viable, and if it relates even partially to the GTS, is it surprising that it is "preserved virtually unchanged within the neo-catastrophic paradigm"? (Tyler)

Do you know of any neo-catastrophic literature that states that since most of the sediments are laid down in catastrophic bursts, all fossils included are not the result of slow evolutionary changes over long periods of time but rather the result of rapid burial of various species all living together at the same time. If you know of such statements, please give me the references.

9. Why do you think the post-Flood boundary should be marked by a global feature? It was dry enough for Noah and the animals to disembark, so should we expect all the land to be drained of water? (Tyler)

Unless one postulates a slow, gradual retreat of the Flood waters (for whatever reason) and the persistence of local regional lake/seas, (with all the complications it brings: lots of sedimentation after the Flood, what eroded the deep surface and especially submarine canyons etc.) one is to expect that the mergence of land/recession of the ocean was a global feature and that should be clearly visible.

seven

The Geological Column Is a General Flood Order with Many Exceptions

Michael Oard

Abstract

Evidence is presented for a middle position on the question of whether the geological column represents an exact sequence of Flood events or not. Local stratigraphic sections seem to line up with the general order of the geological column at hundreds of locations around the world. But there are many problems with the details, such as 1) the geological column is a vertical or stratigraphic representation abstracted from rock units that are mainly found laterally adjacent to each other in the field, 2) research continues to expand fossil stratigraphic ranges, 3) different names are given to the same or a similar organism when found in different-aged strata, 4) taxonomic manipulation, 5) anomalous fossils, and 6) out-of-order fossils. Walker's model, based solely on the Bible, is advocated for analyzing the rock record and is based on the true mechanism for the deposition of the strata. Comparing Walker's model to the geological column reveals several surprises. First, sedimentary rocks labeled Precambrian (if from the Flood), Paleozoic, and Mesozoic strata are early Flood. Second, Cenozoic strata can be early Flood, late Flood or post-Flood depending upon the location and the particular fossil used to define the Cenozoic. Third, Flood deposition is highly nonlinear with a large percentage of strata deposited early in the Flood.

Introduction

The subject of how the geological column fits into Flood geology and the order of events before, during, and after the Flood is quite controversial within creationism. Some creationists advocate that the geological column is an exact representation of the events of the Flood and possibly post-Flood deposition, minus the uniformitarian timescale. In other words, the Cambrian is early in the Flood, followed by the Ordovician, etc., all over the world according to the exact order of the geological column. In that scheme, Mesozoic would be considered middle Flood or late Flood, depending upon where one places the Flood/post-Flood boundary, and the Cenozoic would be either late Flood or post-Flood. Other creationists want to throw out the entire geological column. Still others believe that it is a general sequence with many exceptions.

I will argue the middle position in this chapter: the geological column is a general Flood sequence with many exceptions. Furthermore, I advocate viewing the rocks and fossils through Flood glasses—through the *actual mechanism* that produced the rocks and fossils, the Genesis Flood. Why look at the rocks and fossils through a false philosophical system based on the hypotheses of uniformitarianism, an old Earth, evolution, and naturalism? Since I believe that the geological column is a general sequence of the Flood, I expect some overlap between a Flood classification and the geological column.

I advocate the model or classification of Walker (1994), which is similar to the model derived by Whitcomb and Morris in *The Genesis Flood* (1961). Although Froede (1995) produced a similar model, I prefer Walker's model mainly because it is more developed with defining criteria for his stages and phases. Klevberg modified Walker's timescale for the stages to correspond with the Flood peaking on Day 150 (Figure 1), which seems to be the Scriptural position and also corresponds to the 21 weeks of prevailing and the 31 weeks of assuaging in the Whitcomb-Morris (1961) model.

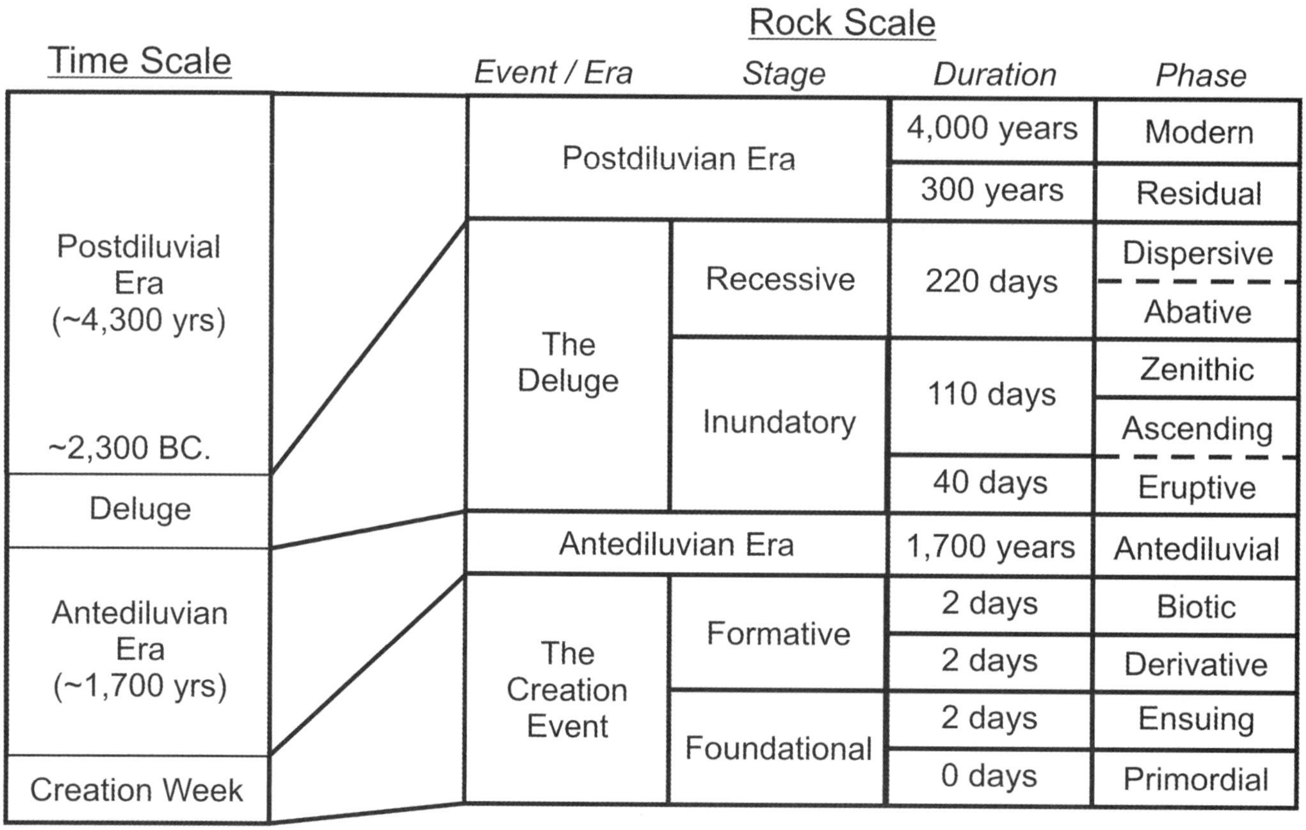

Figure 1. Walker's biblical geological model, modified by Klevberg.

Is the Geological Column a Global Sequence?

Secular scientists often claim that the fossil order represented by the geological column is an "absolute" global sequence. Is this true or just taken on faith?

In order to demonstrate that the geological column is a global sequence, four steps are necessary: (1) develop local columns for small areas, (2) tie local columns into a regional- or subcontinental-scale column, (3) integrate local and regional columns into a continental-scale column and (4) develop the overreaching global geological column. Presumably the first and second steps could be fairly straightforward, if the geology is uncomplicated and the lithology of the strata can be traced for long distances. But, in areas of tectonics, overthrusts (to be discussed below), and facies changes, the development of even a local column may be difficult or nearly impossible. But, in areas of tectonics, overthrusts (to be discussed below), and facies changes, the development of even a local column may be difficult. The third and fourth steps become much more difficult since lithologies and fossils cannot be traced across continents and from continent to continent. It would seem that the task grows by orders of magnitude at these last two stages, becoming more hypothetical the greater the area of extrapolation. Woodmorappe (1999b, p. 24) noted:

> As one moves from local all the way to global correlation by fossils, correlations become increasingly less empirical and more conceptual. This is because there are progressively greater differences (such as lithology, local fossil succession, and overall faunal character) as one moves even further geographically from a reference section in the type area.

The geological column was first developed at a local or regional scale before it was extrapolated to a global scale. The geological column was first set up in England, the Alps, and the Ural Mountains of Russia based on a number of assumptions (see chapter 2). It is possible that the formations in England may be well-behaved vertically and horizontally (this should be checked), so that the part of the column developed in England may be generally accurate. I question how well the Alps and the Permian from the Ural Mountains originally fit into the original geological column because of their distance from England.

Although evolution is claimed to have not been a guiding principle for the construction of the geological column in the early 1800s, the formations were nonetheless pigeonholed into slots based on *fossil succession*. In other words, the original column was not necessarily developed from lithology but mainly by a succession of index fossils. Index fossils are organisms that are assumed to have spread over much of the world and lived only a short time. "Catastrophists" generally developed the column, but these catastrophists believed in multiple catastrophes in which the Genesis Flood was just the last and accounted for only the surficial "diluvium." Some of these catastrophists would be considered progressive creationists today, but others succumbed to uniformitarianism. Fossil succession over long periods of time was the guiding principle, which essentially is the same as evolution. When biological evolution came on the scene, the fossils' succession became the evolutionary progression with time. As it later turned out, much of the "diluvium" was the result of glaciation. So, the Genesis Flood, after first being relegated to the surficial layer, was then rejected entirely by most scientists in the 1800s. Some scientists and theologians held on to a local or tranquil flood, although Scripture is abundantly clear that the Flood was catastrophic and covered the entire Earth.

Many people believe index fossils were supplemented by radiometric dating in the 1900s, but index fossils continue to have preeminence in dating. Radiometric dates *must agree* with the geological column, or the radiometric dates are assumed wrong for various reasons (McKee, 1972, pp. 24–30). As a result of this circular reasoning, there are countless problems in radiometric dating (Woodmorappe, 1999a; Vardiman et al., 2000). A new creationist research project, called RATE (Radioisotopes and the Age of the Earth), has shown that in some instances the millions or billions of years are very likely the result of accelerated radiometric decay on a young Earth (Vardiman et al., 2005).

Even if the fossil succession is more or less accurate for England, the question of the validity of the geological column really boils down to how well the original fossil order from England represents a worldwide order? This question must be answered empirically. The literature indicates that a general order seems to exist but problems occur in the details. This does not imply that an evolutionary order exists, but it is a burial sequence during the Genesis Flood.

Local Columns Show General Order

The justification for the global column is that the small number of index fossils in any one area still line up vertically in their expected order. Of course, creationists should verify this vertical order, especially in view of the problems discussed below.

Trilobites and dinosaurs, organisms from different environments, illustrate the concept of a vertical fossil relationship. If every outcrop shows dinosaurs always superimposed above trilobites, we can have general confidence that this relationship holds as a worldwide relationship in the Flood. Furthermore, if we find a region with just trilobites, we can surmise that the strata were laid down earlier than strata containing dinosaurs in another region. Because of the many problems listed below, there may be exceptions. So, in this case dinosaurs above trilobites would be considered a general Flood order.

Dinosaurs and trilobites lived in quite different environments, and we would expect a vertical order in the Flood. However, I would be more cautious in developing a vertical order with organisms from the *same or similar environments*, such as various types of trilobites, cephalopods, foraminifera, diatoms, etc. They mostly live in a marine environment and during the Flood could have become vertically superimposed in any order, unless there were other factors that could cause systematic vertical relationships, such as ecological zonation, horizontal separation, etc.

The general order of the geological column (Paleozoic below Mesozoic below Cenozoic) seems to be correct on a broad scale in north central Wyoming and south central Montana. Paleozoic strata with trilobites, brachiopods, etc. and Mesozoic strata with dinosaur fossils are commonly found in the mountains, while Cenozoic strata with fossil mammals predominantly occupy the basins and valleys. Paleozoic and Mesozoic strata are often tilted at a high angle at a basin edge against granite intrusions and uplifts of sedimentary rocks in the northern Rocky Mountains (Figure 2), while the Cenozoic strata are nearly flat-lying in the center of the basins. The uplifted Bighorn and Beartooth Mountains and the Bighorn Basin in between are a good example. The Cenozoic strata of the Bighorn Basin and the adjacent Clarks Fork Basin to the north are well known for their fossil mammals. These Cenozoic basin fills postdate the strata in the surrounding mountains. Assuming that the Paleozoic and Mesozoic have typical index fossils for those periods, the order of the fossils lines up with the geological column in this area.

Another example is the Grand Staircase in northern Arizona and southern Utah. Although the Grand Staircase is both a vertical and horizontal relationship, in that the Mesozoic strata lie predominantly to the north of the exposed Paleozoic strata of Grand Canyon, there is strong evidence that the Mesozoic

Figure 2. Tilted Paleozoic and Mesozoic strata at the northwest edge of the Bighorn Basin at Clark Canyon adjacent to the southeast Beartooth Mountains.

strata once lay above the Paleozoic Grand Canyon. The Mesozoic strata were later eroded, leaving remnants such as 300 m-high Red Butte along the southeast rim of the Grand Canyon (Figure 3). However, I would question the Cenozoic age of the Wasatch Formation on top of the Mesozoic section in Utah. I believe this formation name was based on the assumption that strata on top of Mesozoic must be early Cenozoic, and since the Wasatch Formation crops out in basins to the north, the top strata likely were simply rubberstamped as the Cenozoic Wasatch Formation. However, the top formation of the Grand Staircase is no longer considered to be the Wasatch Formation; it is the Clarion Formation (Harris et al., 1997, pp. 43–54). However, the Clarion Formation is still considered to be Cenozoic. The unique erosional forms of Bryce Canyon were carved in the Clarion Formation (Figure 4). Fossils in the Clarion Formation are not abundant (Harris et al., 1997, pp. 52–53), so it is unlikely that fossils can be used to determine its age. If the formation was actually "Mesozoic," then only two of the three Phanerozoic eras of the geological column are represented in the Grand Staircase.

I question the finer time divisions within the Paleozoic or Mesozoic, such as the division between the Cambrian, Ordovician, Silurian, etc. The Paleozoic commonly contains marine deposits (one exception being the claim that the Coconino sandstone is a desert deposit, which is debatable). The environmental interpretation is based on marine fossils such

Figure 3. The erosional remnant of Red Butte near the south rim of the Grand Canyon (view south from the north rim). The butte is 300 m above the planation surface at the top of the Grand Canyon.

Figure 4. Unique erosional forms in Clarion Formation of Bryce Canyon National Park.

Figure 5. Trilobite tracks from the Grand Canyon (arrows).

as trilobite tracks (Figure 5) and nautiloids (Figure 6) found in the Grand Canyon and at other locations. It is likely these organisms lived before the Flood, and so the Paleozoic represents a marine burial sequence, possibly by ecological zonation. Between the Cambrian Muav Limestone and the Devonian Temple Butte Limestone, the Ordovician and Silurian periods, with their 120 million years of geological time, are missing. The contact between the Muav and Temple Butte is a disconformity, a break in deposition or an erosional event between parallel beds. Figure 7 shows a fold at the disconformity, implying little if any time gap, because the lower limestone formation should have already been lithified and thus could not have been folded parallel to the upper formation. If the geological column is an exact Flood sequence, this disconformity would represent a period of erosion or nondeposition between the Muav and Temple Butte Limestones during the Flood. However, if the geological column is merely a general order, there is no reason to suggest a period of nondeposition or ero-

Figure 6. Nautiloid from the Grand Canyon.

Figure 7. Disconformity between the Muav Limestone and Temple Butte Limestone in the Grand Canyon. Notice how folding affects both formations.

sion between the two limestones. The specific index fossils for those periods simply were not deposited. I might add that the Ordovician and Silurian are also considered absent in practically all of Montana (Perry, 1962, p. 24), likely because of missing index fossils. If someone found an index fossil for the Ordovician, you can be sure that strata now labeled Cambrian or Devonian would become Ordovician or Silurian.

Reed (2005) advocated that creationists with geological knowledge become familiar with the geology and paleontology of their local area for eventual regional scale investigation. We can focus just on the rock record and develop our own local geological columns. In this way we would be able to analyze the rock record from each local area and relate it to a global Flood model.

Figure 8. Part of Wyodak coal seam just east of Gillette, Wyoming.

Problems for the Geological Column

Despite propaganda by evolutionary and uniformitarian scientists that the fossil order is an exact global order with time, there are numerous problems and anomalies that make this assertion questionable. I can only briefly mention these problems, since they could be amplified into a whole book.

1) Vertical Sequence of Geological Column is often Horizontal in the Field

Many think that the geological column is a vertical, onion-skinned model, which has the same vertical sequence in most areas. Actually, the vertical fossil scheme is mostly derived from lateral relationships. The reason for this is because only a small number of the ten Phanerozoic geological periods are represented as a vertical sequence in any local area, defined as a 406 by 406 km square (Woodmorappe, 1999d). Two-thirds of Earth's land surface has five or fewer of the ten geological periods in place. Only 15–20% of Earth's land surface has even three geological periods in correct order. This is a conservative estimate in favor of the geological column because Woodmorappe used any suggestion of a period being in a square as evidence that the period existed in that particular square. His squares are so large that it was difficult to establish a single vertical sequence because of tectonics, facies changes, etc., and many of these local geological columns should be verified lithologically. Regardless, the global and continental columns mainly represent a horizontal sequence. Unless there are better empirical correlations, it may be difficult to know the exact time sequence in the Flood over such large areas.

For instance, the late Paleozoic is well represented by coal from trees such as lycopods in the Appalachian Mountains, while in Montana and Wyoming the coal (Figure 8) contains angiosperms and gymnosperms. The coal in Montana and Wyoming is dated as "early Cenozoic," much younger than the Appalachian lycopods in the geological column. But, the different trees really represent a horizontal separation. Whether or not the different plants making up the coals represent an age sequence in the Flood must be determined empirically.

The horizontal relationship of index fossils is also a global phenomenon (Woodmorappe, 1999b). In a study of 34 index fossils, Woodmorappe found that only rarely are more than a third and never more than a half of these index fossils simultaneously present in any 320 km-diameter region on Earth. And even those index fossils found in a particular region are rarely vertically superimposed.

The problem is that it is doubtful enough of these local relationships can be traced horizontally to know whether the global geological column really represents a vertical sequence. For example, the coals from the Appalachians and from the Montana/Wyoming area could have been laid down at the same time in the Flood.

So, the global geological column is built by extrapolating periods and index fossils from each area into a global sequence. How well this global sequence lines up with reality and represents a Flood order requires much more research, but I am skeptical that each period in the geological column represents an absolute sequence of events in the Flood model.

2) Changing Fossil Ranges in the Geological Column

In order to discuss fossil order, we need to know the three-dimensional distribution of fossils. Fossils come from scattered outcrops and boreholes; we know very little of the subsurface distribution of fossils. The more scientists examine the rocks, the more the *ranges of fossils are extended* in the geological column (Woodmorappe, 1999e, pp. 135–136).

For instance, organisms thought to have been extinct for millions of years sometimes are found alive in remote locations on Earth. These organisms are called living fossils. Logically, these organisms must have lived during later geological periods where their fossils have not been discovered. If this applies to many other organisms, fossil ranges for many organisms can be greatly extended *upward* toward the present.

One of the most recent outstanding examples of a living fossil is the Wollemi Pine (Figure 9), found in a gorge in the Blue Mountains, 200 kilometers west of Sydney, Australia (Wieland, 1995). The Wollemi Pine was thought extinct since the Jurassic period—about 150 million years ago on the uniformitarian timescale. This means that the Wollemi Pine should

Figure 9. Wollemi Pine from Blue Mountains of New South Wales (courtesy of Creation Ministries International, Australia).

exist in strata between the Jurassic and the present. One researcher described the discovery like "finding a live dinosaur" (Wieland, 1995). Obviously, no evolution of the Wollemi pine has occurred for an alleged 150 million years. Given its absence in strata younger than "Jurassic," those 150 million years may never have existed. One would expect abundant Wollemi pine fossils during this 150 million-year period. Catastrophic burial about 4,500 years ago is a better explanation for living fossils, such as the Wollemi pine.

A sponge, called *Nucha? vancouverensis sp. nov.*, was found in the upper Triassic of Vancouver Island (Stanley, 1998). Surprisingly, this sponge is nearly identical to one previously found only in the Middle Cambrian of western New South Wales, Australia, which was named *Nucha naucum* (Oard, 2000). The fossil has not been found in strata within the 300 million intervening years. Assuming that the paleontological analysis on these sponges is correct, the range of *Nucha* is significantly expanded upward in the geological column, and one wonders whether the 300 million years between the Cambrian and the Permian are real. The above situation is not rare (Stanley, 1998). These examples should make us aware that paleontologists do not know the three-dimensional distribution of fossils, and that the many millions of years between the same or similar fossils may not exist.

Fossil ranges have also been extended *downward* in the geological column. For instance, vertebrates have been pushed back into the Cambrian (Oard, 1996a; 2004a) where 50% to possibly as high as 85% of all phyla originated in what is now called the Cambrian Big Bang (Meyer et al., 2003). Sharks have been pushed back 25 million years into the Late Ordovician (Oard, 1996a). Vascular plants have also been pushed back 25 million years into the Early Silurian (Oard, 1996a). Based on tracks, arthropods invaded the land 40 million years earlier (Late Cambrian) than previously thought (MacNaughton et al., 2002; Oard, 2003). The discovery of a possible winged insect would push back the origin of winged insects and flight by more than 80 million years into the early Silurian, which in turn has caused the supposed first land plants to be pushed back into the Ordovician (Engel and Grimaldi, 2004; Oard, 2004b).

If their analysis of organic molecules is correct, evolutionists believe that they have pushed back the origin of eukaryote cells one billion years to 2.7 billion years ago in the late Archean (Brocks et al., 1999; Oard, 2001a). This raises interesting questions for both evolutionists and creationists. Where are the remains of all the billions of organisms with eukaryote cells that lived between 2.7 billion years ago and the time of the Cambrian Big Bang (~500 million years ago) in the evolutionary model? Since the molecules were found in sedimentary rocks, does this mean that Archean and Proterozoic sedimentary rocks are from the Flood?

3) Different Names for the Same or Similar Fossil from Different Ages

It is not an uncommon phenomenon to find the same or similar fossils in strata of different ages that have been given *different names*. Very few non-specialists would be aware of this phenomenon. This practice masks the true range of fossils within the geological column. Tosk (1988) documented that the same or similar foraminifera are not only given different names when found in strata of different ages, but also are sometimes placed in different superfamilies. Woodmorappe (1999f) found that much of the stratigraphic order of cephalopods is due to time-stratigraphic concepts and taxonomic manipulation. Both cephalopods and foraminifera are important index fossils.

The same situation occurs with plants. Rees et al. (2000, p. 301) complain:

> Indeed, it is sometimes necessary to 'side-step' traditional paleobotanical taxonomy, which

is often hindered by political and regional biases (ensuring a highly specialized local but limited global view), as well as stratigraphic biases (with what is effectively the 'same' fossil plant type being assigned to a different genus or species depending upon its age).

4) Taxonomic Manipulation

Another problem mentioned by Woodmorappe (1999f) is that slightly different features in cephalopods have been used to date a layer of strata to a different age. These slightly different biological features cause one type of organism to be split into a different species, genera, families, etc. Since taxonomic splitters have had the upper hand in taxonomy, how meaningful are such taxonomic and age manipulations to the geological column? We know that species of living organisms, like dogs and pigeons, have a great morphological variety. How do we know whether the variety found in an extinct organism is not from intraspecies variation? Within creationist biological terms, such variation would be considered within the same Genesis kind or *baramin*.

For example, one type of trilobite might date a layer as Cambrian while a slight change in anatomy in another trilobite in another layer will cause that particular layer to be dated as Silurian. Are they different kinds of trilobites or variations within one kind?

These problems make it difficult to take seriously the separation of the periods within the Paleozoic and Mesozoic. The Paleozoic may simply represent mostly marine deposition during the Flood. Trilobites buried at nearly the same time are assigned from the Cambrian to the Permian in the uniformitarian system. On the other hand, the organisms of the Mesozoic are much different, and generally above Paleozoic fossils where they are found vertically superimposed. So, the order of the geological column seems like a general sequence from a Flood depositional point of view, but with lots of exceptions in the details.

5) Anomalous Fossils

Evolutionists often tell us that there are no contradictions to the evolutionary fossil order. However, they have to explain many anomalies in order to make the geological column "consistent." One type of anomaly is finding two fossils of different ages in the same layer.

If the evolutionist cannot extend the stratigraphic range of the fossils, he must determine which fossil represents the true "age." If the strata are considered young, the "old" fossil is simply assumed to have been "reworked," eroded from "much older" strata and incorporated into younger sediments. Often, their only criterion for reworking is an expected evolutionary order. If "old" organisms are reworked into "young" strata, wouldn't the "old" fossil be pulverized?

In the opposite case, a "young" fossil is found in "old" strata, and evolutionists assume that the "younger" organism was buried within "old" sediment and fossilized. This is called "downwash." This could happen if a "young" organism became trapped and fossilized in a cave, sinkhole, or bog within "old" sediment or sedimentary rock. If the strata remain unconsolidated until after the "young" organism is buried, it would be difficult for the "old" organism to have remain unfossilized for millions of years.

Whether a fossil is considered reworked or downwashed, there should be evidence for such an event.

Woodmorappe (1999c, pp. 87–92) compiled 200 published instances of anomalous fossils from the literature. This was not an exhaustive search. Most of these instances involved microfossils, which is why I am especially skeptical of the biostratigraphy of various microfossil groups, such as foraminifera and diatoms. Taxonomic manipulation, along with reworking, casts doubt on the use of microfossils as index fossils. Anomalous fossil occurrences are not rare (Woodmorappe, 1999c, pp. 92–94). Furthermore, if evolutionists underreport examples of anomalous fossils, they may be quite common, while evidence for reworking or downwash is rare! It seems that reworking is just an *ad hoc* explanation to make the geological column "consistent." The real impact of anomalous fossils would be to broaden the fossil range in the geological column, thereby reducing confidence in index fossils.

6) Out-of-Order Fossils

A second type of anomaly in the fossil record is the situation in which "older" fossils are found above rocks that contain "young" fossils. These out-of-order fossils are the opposite of the evolutionary hypothesis. Out-of-order fossils are considered "impossible" by evolutionists, and so are dismissed as the result of *overthrusting*. An overthrust involves "older" strata being pushed over "younger" strata at an angle less than 45°.

Robinson (1996, p. 35) claimed that overthrusts are based on geophysical evidence and not out-of-order fossils. This is true for some, but the Lewis overthrust in Montana and Alberta (Figures 10–12) was identified based on fossils. In the Lewis "overthrust," Precambrian rocks supposedly slid tens of kilometers eastward up a low slope over "Cretaceous" rocks. There is a 900 million-year out-of-order time gap

Figure 10. Lewis "overthrust" northeast of Marias Pass, Montana (view northeast). The "Precambrian" Altyn Dolomite is the light colored layer in the center of the picture while the Appekunny argillite is the dark colored rock above. "Cretaceous" shale lies below the dolomite. Note the horizontal beds of the shale, which are either undeformed or only mildly deformed below the contact.

Figure 11. The contact of the Lewis "overthrust" northeast of Marias Pass.

Figure 12. Close-up of the contact of the Lewis "overthrust" northeast of Marias Pass. There are stringers of Altyn dolomite in shale below contact.

at the Lewis "overthrust," and this time gap was first based on out-of-order fossils. Bailey Willis (1902), who first hypothesized the "overthrust," found "Precambrian crustacean shells" in the upper block above the "Cretaceous" strata. The Lewis Overthrust may or may not be a true overthrust, but the determination should be made by geological and geophysical methods and not by fossils.

Another famous example of an overthrust is the Heart Mountain detachment in north central Wyoming. It is not a true overthrust but the upper block actually slid down a slight incline and broke up into many smaller blocks. That is why it is now called a detachment fault. Heart Mountain north of Cody, Wyoming, is the most famous example (Figure 13). The Heart Mountain Detachment is real and there is evidence for motion, such as broken rock at the detachment surface (Beutner and Gerbi, 2005). So in this case, there is a structural explanation for the out-of-order fossils.

A modern analog for the Heart Mountain Detachment (Oard, 1996b) was discovered when large blocks of lava detached from Hawaii and slid into the deep ocean (Moore et al., 1995). In the South Kona Landslide, one huge block broke up into large pieces, up to 700 m high and 11.5 by 7.5 km in area. It slid up to 80 km oceanward—the last 40 km over relatively *flat* ocean bottom. These blocks are larger than the Heart Mountain Detachment blocks. Most uniformitarian geologists believe that the Heart Mountain Detachment was catastrophic, occurring within a matter of minutes or hours (Beutner and Gerbi, 2005).

In such cases, there is evidence of overthrusting or reverse faulting. A reverse fault is the case where a block is shoved up over other rock at an angle greater than 45°. I believe that there is evidence of thick-skinned reverse faults and even overthrusts. For instance, in some regions of the Bighorn and northeast Beartooth Mountains of south central Montana and north central Wyoming, granite has been pushed east or northeast up an approximately 30° slope (Wise, 2000; Stone, 2003). Such thick-skinned (granite is involved) overthrusts are supported by seismic profiles and wells drilled on the eastern edge of the granite that pass into sedimentary rock.

The fault zone of the Beartooth thrust consists of 21 m of shattered granite above 37 m of severely faulted sedimentary rocks (Wise, 2000, p. 366). Such evidence should also exist with thin-skinned "overthrusts," in which sedimentary rock is pushed over

Figure 13. Heart Mountain, northwest Bighorn Basin. The light colored strata at the top of Heart Mountain are "Paleozoic" limestone and dolomite, which lies on top of valley fill sediments (view south southeast)

sedimentary rock. However, I have seen a number of overthrusts in Montana and southern Alberta where there is usually little or no evidence for displacements of km to tens of km uphill over a slope less than 45° (Woodmorappe, 1999c, pp. 86–87).

Some "overthrusts" display a reversed metamorphic grade in which the upper block is more highly metamorphosed than the lower block. Metamorphism is supposed to increase with increasing depth. So, this is support for the overthrust concept in these cases. However, it is possible that the metamorphic grade associated with "overthrusts" could be chemically caused (Silvestru, personal communication) or caused by the migration of heat and fluids during deformation (Hubbard, 1996). Overthrusts, if they are real, could possibly be explained by catastrophic underwater emplacements during the Flood. Creationists need a comprehensive analysis of overthrusts.

When one realizes that there are hundreds of alleged overthrusts (they seem to occur in most mountain ranges of the world), and that mountains are usually the few places to observe a thick vertical sequence, one is forced to conclude that out-of-order strata are common. A real overthrust should show abundant physical evidence; relying just on fossils is unreasonable. If these strata cannot be tied to a real overthrust, then the fossil distribution in the geological column is contrary to evolutionary predictions.

Does the Geological Column Represent the Flood Depositional Sequence?

These problems should make any creationist cautious in applying the geological column to the Flood. We seem to be divided by those who believe that the geological column is an exact sequence of the Flood and those who want to entirely discard the column. After many years, I have come to the conclusion that the geological column represents the general order of the Flood with many exceptions, and that its application should be made with caution for the reasons enumerated above.

In examining fossils and fossil successions with regard to the Flood, we must distinguish between animals that survived the Flood and those that did not. This distinction will help determine whether a fossil was buried by the Flood or is post-Flood. The animals that God brought onboard the Ark were a male and female of each unclean kind and seven of each clean kind. These animals had to be terrestrial and breath air (Genesis 7:21, 22).

The Genesis *kind* cannot be equated with modern species in many cases (Woodmorappe, 1999e, p. 136). If the kind is at the genus level, the ark needed only 16,000 animals (Woodmorappe, 1996), primarily mammals, birds, and reptiles. Many other organisms could have survived the Flood outside the Ark. Therefore, all mammals, reptiles (including dinosaurs), and likely all birds had to be dead by the time the water started retreating off the land around Day 150 (Genesis 7:22–8:3). So, evidence of a live mammal or reptile would indicate either an early Flood or post-Flood time. Marine organisms, such as foraminifera, could potentially represent early Flood, late Flood, or post Flood.

1) Walker's Model

To bypass all the confusion with the geological column, I advocate Walker's (1994) model of the Flood (Figure 1). Viewing the strata through flawed uniformitarian concepts does not seem logical. So, we need to put on our "Flood glasses" when looking at the rocks and fossils. Walker's model was derived directly from the Bible apart from the geological column or any other philosophical presupposition. It also provides a template for examining how the geological column relates to the Flood.

Walker's (1994) model is at odds with even the relative dating of the column. For example, Walker (1996a) classified the basement rocks around the Brisbane area as being from the Eruptive Phase of the Inundatory Stage of the Flood—its very beginning, even though these rocks are generally dated as middle Paleozoic in the geological column. Walker (1996b) then assigned the shale and sandstone deposits of the Great Artesian Basin to the upper Zenithic Phase of the Inundatory Stage (just before the floodwater peaked). The strata of this basin cover an area of 1,800,000 km^2 and are over 2,000 m in thickness. They are dated as mostly Jurassic and Cretaceous in the geological column, but represent the first half of the Flood. Thus, in eastern Australia, the Paleozoic and Mesozoic strata are *early* Flood.

2) Precambrian to Mesozoic Strata in the Rocky Mountains

In the Rocky Mountain region of the United States, Precambrian sedimentary rocks commonly outcrop in mountain ranges and their thickness indicates that they represent deposits from large, isolated basins that have uplifted. Examples include the Belt Supergroup that forms the northern Rockies of western Montana and northern and central Idaho, the Uinta Mountains in northeast Utah, and the Precambrian sedimentary rocks in the eastern Grand Canyon. Whether these Precambrian sedimentary rocks are pre-Flood or Flood has not yet been resolved.

Paleozoic and Mesozoic strata can form large sheets over extensive areas such as the Great Plains, but they are generally broken and tilted in the mountains in the western United States, except for the Colorado Plateau. It is possible that the Paleozoic and Mesozoic strata in the Rocky Mountains were once continuous over the region like on the Colorado Plateau.

Tracks are one of Walker's (1994, p. 589) defining criteria for the Inundatory Stage. The Mesozoic of the Rocky Mountains and High Plains has millions of dinosaur tracks, as well as thousands of eggs, on flat bedding planes. It seems obvious that these tracks and eggs are from the Flood, and since they represent live dinosaurs, the Mesozoic in this area would be from the Inundatory Stage, early in the Flood (Oard, 2004d, pp. 103–105). So, these Paleozoic and Mesozoic strata were deposited early in the Flood, similar to eastern Australia. Although the general sequence of Paleozoic to Mesozoic seems valid, the periods within those eras may not represent an exact sequence, since the Devonian in one place may be deposited before the Cambrian in another.

3) The "Cenozoic" can be Early Flood, Late Flood, or Post Flood

The "Cenozoic," on the other hand, is the most problematic (Oard, 2001c, pp. 89–90). It generally fills basins in the Rocky Mountains and outcrops as sheets on the High Plains. There are indications of erosion of many hundreds and even a few thousand meters of rock in these areas (Oard, 1996, 2001b, Oard and Klevberg, 2005). The high areas of the western United States are a scoured surface. That is why there is so much bedrock close to the surface in those areas. There is clear evidence for sheet erosion followed by channelized erosion, which correspond to Walker's two phases of the Recessional Stage of the Flood (figure 1). This erosion must have occurred mainly in the Recessional Stage of the Flood between Days 150 and 371. So, much of the Cenozoic strata not eroded in the Rocky Mountain basins and High Plains were likely deposited during the Inundatory Stage of the Flood. Some of this strata is dated late Cenozoic in the geological column (Thompson et al., 1982), implying that "late Cenozoic" strata can be early Flood!

There also are mammal tracks in some of the Cenozoic strata in these basins that reinforce the deduction that most of the remaining Cenozoic strata were deposited in the Inundatory Stage (Lockley and Hunt, 1995, pp. 243–281; Oard, 1998, pp. 79–81). Based on Walker's (1994) model, tracks of mammals on Flood strata must have occurred in the Inundatory Stage. This evidence indicates that practically all strata, clear up to the Pliocene, in the higher areas of the western United States were deposited early in the Flood during the Inundatory Stage.

Sediments eroded from the high areas of the western United States were redeposited far to the west and east. Eroded debris would have been deposited in deeper areas where currents would decrease. Strong currents eroding the uplifting western United States would have pulverized much of the rock, but the most resistant rocks would have been carried far from the source and deposited as a lag or as basin fill. The most resistant rock of significant volume is quartzite. Quartzite cobbles and boulders, well rounded by water, are found over 1000 km to the east and 700 km to the west of their Rocky Mountain sources (Oard et al., 2005, in press a; in press b). These quartzites are practically all dated as Cenozoic by mainstream geologists, based on included mammal fossils, especially in interbeds, but they would be part of the Recessional or late Stage of the Flood.

Furthermore, the eroded strata would have been redeposited on the continental shelf off the western U.S.—a Recessional Stage feature of the Flood (Walker, 1994; Spencer and Oard, 2004). The eroded material probably would also have been deposited in basins near the coast, such as the lower Mississippi River Valley. Much of the Cenozoic strata of Washington, Oregon, and California could be Recessional Stage sedimentation. Mammals, which are found in Cenozoic high western U.S. basins, should be mostly pulverized by the powerful recessional stage currents and turbulence, which Klevberg and Oard (1998) estimated would have flowed over 30 m/sec. The strata in these areas are generally dated as "Cenozoic" by microorganisms and terrestrial mammals. These Cenozoic strata would be a late Flood or Recessional Stage feature.

Massive Recessional Stage erosion may also explain sparse human fossils in sedimentary rocks. If human remains were mostly deposited in the upper sedimentary layers by Day 150, these layers would have been heavily eroded from currently high areas of Earth, pulverized, and deposited over lower areas (Austin et al., 1994, p. 614).

There is also the likelihood that some "Cenozoic" sediment on the bottom of the ocean, mostly dated by microfossils, is post-Flood, although microfossils could have been laid down early in the Flood, late in the Flood, or afterwards. Microorganisms would have proliferated in the oceans during the Recessional Stage of the Flood because of the huge amount of nutrients flowing into the ocean and mixing at all depths. High microorganism productivity would be expected to continue after the Flood due to the warm ocean and rapid overturning during the Ice Age that would help keep nutrient levels abundant in the upper layers of the ocean (Oard, 1990, pp. 70–75). The Flood probably deposited the deeper sediments while the upper sediments are likely post-Flood, although ocean bottom reworking (Thiede, 1981; Woodmorappe, 1999c, pp. 87–93) would result in exceptions. Some Paleocene ocean bottom sediments may be post-Flood, while some Pliocene sediments could be from the Flood, based on uncertainties in evolutionary microorganism classification.

Another indicator of post-Flood Cenozoic sediments on the bottom of the ocean is ice-rafted material. Ice rafting into the ocean would be expected in the middle to late Ice Age because sufficient time is required for glaciers and ice sheets to build and spread to the oceans, which were warm at the beginning of the Ice Age (Oard, 1990; 2004c). Ice rafted debris (if the interpretation is correct) is found in sediment dated by microfossils as Oligocene and Miocene (Oard, 1998, p. 81). Some of the sediment from the early Ice Age could be dated as Paleocene or Eocene by uniformitarians. If the oxygen isotope/temperature relation holds generally true for ocean bottom microorganisms, much of the Cenozoic shows a cooling trend, as would be expected in the oceans during the post-Flood Ice Age (Vardiman, 1996).

So, in the Flood model "Cenozoic" can be early Flood, late Flood or post-Flood, depending upon the location. This comparison is based on logical deductions from Walker's biblical geological model and the post-Flood Ice Age. The "Cenozoic," as a worldwide part of the geological column, can refer to almost any specific time in the Flood.

4) Nonlinear Flood Deposition

Many creationists have assumed a linear relationship between the geological column and the Flood and post-Flood period with the Cenozoic being late Flood or post-Flood (Garner et al., 1996). However, based on Walker's model and reasonable defining criteria for his stages and phases, Flood deposition appears highly nonlinear with respect to the geological column. Practically all the current strata in the high western United States (and probably some of that eroded) were deposited *early* in the Flood. It is highly unlikely that "Cenozoic" strata in the high western United States are post-Flood or even late Flood (Oard, 1996b; Oard and Whitmore, 2006; Oard, in prep; Oard and Klevberg, 2005). Thus, a vast amount of deposition occurred in the western United States early in the Flood. This has serious implications for any Flood model. Most creationist believe that the most violent part of the Flood was the beginning with the start of the catastrophic mechanism, while the later half of the Flood was more subdued and mainly an erosional event caused by differential up or down motion of the crust and upper mantle (Oard, 2001b; 2001c). This generally goes along with the geological energy curve of Reed et al. (1996).

Summary

In this paper I have argued for a middle position between the creationist views of accepting the geological column as an exact, linear chronology of the Flood and those that want to dismiss the column altogether. I first outlined what must be accomplished in order to show that the geological column is an exact sequence for either the uniformitarian or Flood paradigm. One must develop local and regional columns that must be shown to have a continental and global consistency. However, more empirical local columns become more theoretical as one extrapolates to larger areas.

Secondly, I showed how the geological column seems to be generally consistent where observed in vertical sections in the western United States. This gives some confidence that the general order can be applied elsewhere in the world. However, the finer divisions of the eras, especially where the environment of the fossils is similar, should be questioned. Creationists should check every detail of the fossils and the lithology of the rocks to verify that the column is general and to see whether the details of the periods and subperiods line up. There are so many uniformitarian assumptions behind the geological column.

The third section provides reasons for skepticism of the details of the geological column. The geological column seems to be less a vertical sequence at any one location and more a broad horizontal sequence, which is based on index fossils from scattered outcrops that likely are difficult to correlate. Research continues to expand fossil ranges in the geological column, making suspect the use of index fossils for dating within short periods of time. Unfortunately, different names are given to the same or a similar fossil found in strata of different ages. Correct taxonomic classification would likely expand the time range of fossils.

There are several other problems that would significantly expand the time range of fossils, if the observed fossil distribution were the only input. These problems are taxonomic manipulations, anomalous fossils, and out-of-order fossils. All these problems are difficult to evaluate and cause an unwarranted reduction in the time range of fossils. A greater time range for index fossils used to date strata would especially make the fine divisions within the geological column questionable.

In the fourth section, I come back to the question of how well the geological column represents a Flood order of deposition. We should really be looking at the fossils and rocks by the *mechanism* that deposited them, and not by a system that was set up assuming the Flood never occurred and that Earth is old. That is why I recommend Walker's (1994) classification or model, which is based on reasonable deductions from Scripture. Walker uses classification criteria for his phases and stages of the Flood. When we use Walker's model, we find that much of the Precambrian, Paleozoic, and Mesozoic strata were laid down in the Inundatory Stage or the first 150 days of the Flood. Furthermore, the Cenozoic strata can be early Flood, late Flood, or post-Flood depending upon what particular index fossil is used and the location. In other words, Flood sedimentation is highly nonlinear with most sediment deposited in the Inundatory Stage. The Recessive Stage represents mainly continental erosion and deposition on the continental margins.

References

CRSQ—*Creation Research Society Quarterly*

TJ—*Technical Journal* (formerly *Creation Ex Nihilo Technical Journal,* now *Journal of Creation)*

Austin, S.A., J.R. Baumgardner, D.R. Humphreys, A.A. Snelling, L. Vardiman, and K.P. Wise, 1994. Catastrophic plate tectonics: a global Flood model of Earth history. In Walsh, R.E. (editor), *Proceedings of the Third International Conference on Creationism* (technical symposium sessions), pp. 609–621. Creation Science Fellowship, Pittsburgh, PA.

Beutner, E.C. and G.P. Gerbi. 2005. Catastrophic emplacement of the Heart Mountain block slide, Wyoming and Montana, USA. *GSA Bulletin* 117:724–735.

Brock, J.J., G.A. Logan, R. Buick, and R.E. Summons. 1999. Archean molecular fossils and the early rise of eukaryotes. *Science* 285:1,033–1,036.

Engel, M.S. and D.A. Grimaldi. 2004. New light shed on the oldest insect. *Nature* 427:627–630.

Froede, C.R., Jr. 1995. A proposal for a creationist geological timescale. *CRSQ* 32:90–94.

Garner, P., S. Robinson, M. Garton, and D. Tyler, 1996. Comments on polar dinosaurs and the Genesis Flood. *CRSQ* 32 (4):232–234.

Harris, A.G., E. Tuttle, and S.D. Tuttle. 1997. *Geology of National Parks* (fifth edition). Kendall/Hunt Publishing Company, Dubuque, IA.

Hubbard, M.S. 1996. Ductile shear as a cause of inverted metamorphism: example from the Nepal Himalaya. *Journal of Geology* 104:493–499.

Klevberg, P. and M.J. Oard. 1998. Paleohydrology of the Cypress Hills Formation and Flaxville gravel. In Walsh, R.E. (editor), *Proceedings of the Fourth International Conference on Creationism* (technical symposium sessions), pp. 361–378. Creation Science Fellowship, Pittsburgh, PA.

Lockley, M. and A.P. Hunt. 1995. *Dinosaur tracks and other fossil footprints in the Western United States*. Columbia University Press, New York.

MacNaughton, R.B., J.M. Cole, R.W. Dalrymple, S.J. Braddy, D.E.G. Briggs, and T.D. Lukie. 2002. First steps on land: arthropod trackways in Cambrian-Ordovician aeolian sandstone, southeastern Ontario, Canada. *Geology* 30:391–394.

McKee, B. 1972. *Cascadia: the geological evolution of the Pacific Northwest*. McGraw-Hill Book Company, New York.

Meyer, S.C., M. Ross, P. Nelson, and P. Chien. 2003. The Cambrian explosion: biology's big bang. In Campbell, J.A. and S.C. Meyer (editors), *Darwin, design, and public education*, pp. 323–402. Michigan State University Press, East Lansing, MI.

Moore, J.G., W.B. Bryan, M.H. Beeson, and W.R. Normark. 1995. Giant blocks in the South Kona Landslide, Hawaii. *Geology* 23:125–128.

Oard, M.J., 1990. *An ice age caused by the Genesis Flood.* Institute for Creation Research, El Cajon, CA.

———. 1996a. Evolution pushed further into the past. *TJ* 10(2):171–172.

———. 1996b. Where is the Flood/post-Flood boundary in the rock record? *TJ* 10(2):258–278.

———. 1996c. Possible analogue for the Heart Mountain Detachment. *TJ* 10(1):3–4.

———. 1998. Dinosaurs in the Flood: a response. *TJ* 12(1):69–86.

———. 2000. How well do paleontologists know fossil distributions? *TJ* 14(1):7–8.

———. 2001a. Supposed eukaryote evolution pushed back one billion years. *TJ* 15(1):4.

———. 2001b. Vertical tectonics and the drainage of Floodwater: a model for the Middle and Late Diluvian Period—Part I. *CRSQ* 38(1):3–17.

———. 2001c. Vertical tectonics and the drainage of Floodwater: a model for the Middle and Late Diluvian Period—Part II. *CRSQ* 38(2):79–95.

———. 2003. Arthropods supposedly invaded land 40 million years earlier. *TJ* 17(2):3–4.

———. 2004a. Origin of vertebrates confirmed in the Early Cambrian. *TJ* 18(1):10–11.

———. 2004b. 'Evolutionary origins' continue to be pushed back in time. *TJ* 18(3):7.

———. 2004c. *Frozen in time: the woolly mammoths, the Ice Age, and the Bible*. Master Books, Green Forest, AR.

———. 2004d. *The Missoula Flood controversy and the Genesis Flood*. Creation Research Society Monograph 13. Creation Research Society Books, Chino Valley, AZ.

———. In press. Defining the Flood/post-Flood boundary in sedimentary rocks. *Journal of Creation*.

Oard, M.J., J. Hergenrather, and P. Klevberg. 2005. Flood transported quartzites—east of the Rocky Mountains. *TJ* 19(3):76–90.

Oard, M.J., J. Hergenrather, and P. Klevberg. In press a. Rounded quartzite boulders in the northwest United States and adjacent Canada—strong evidence for the Genesis Flood: Part II—long distance transport west. *Journal of Creation*.

Oard, M.J., J. Hergenrather, and P. Klevberg. In press b. Rounded quartzite boulders in the northwest United States and adjacent Canada—strong evidence for the Genesis Flood: Part III—uniformitarian and diluvial interpretations. *Journal of Creation*.

Oard, M.J. and P. Klevberg. 2005. Deposits remaining from the Genesis Flood: rim gravels of Arizona. *CRSQ* 42 (1):1–17.

Oard, M.J. and J.H Whitmore. 2006. The Green River

Formation of the west-central United States: Flood or post-Flood. *Journal of Creation* 10(1):45–85.

Perry, E.S. 1962. *Montana in the geologic past*. Montana Bureau of Mines and Geology Bulletin 26, Butte, MT.

Reed, J.K. 2005. Strategic stratigraphy: reclaiming the rock record! *TJ* 19(2):119–127.

Reed, J.K., C.R. Froede, Jr., and C.B. Bennett, 1996. A biblical Christian framework for Earth history research: Part IV—the role of geologic energy in interpreting the stratigraphic record. *CRSQ* 33(2):97–101.

Rees, P. MCA., A.M. Ziegler, and P.J. Valdes. 2000. Jurassic phytogeography and climates: new data and model comparisons. In Huber, B.T., K.G. MacLeod, and S.L. Wing (editors), *Warm climates in Earth history*, pp. 297–318. Cambridge University Press, London.

Robinson, S.J. 1996. Can Flood geology explain the fossil record? *TJ* 10(1):32–69.

Spencer, W.R. and M.J. Oard. 2004. The Chesapeake Bay impact and Noah's Flood. *CRSQ* 41(3):206–215.

Stanley, G.D. 1998. Triassic sponge from Vancouver Island: possible holdover from the Cambrian. *Canadian Journal of Earth Sciences* 35:1,037–1,043.

Stone, D.S. 2003. New interpretations of the Piney Creek thrust and associated Granite Ridge tear fault, northeastern Bighorn Mountains, Wyoming. Rocky Mountain *Geology* 38(2):205–235.

Thiede, J. 1981. Reworking in Upper Mesozoic and Cenozoic central Pacific deep-sea sediments. *Nature* 289:667–670.

Thompson, G.R., R.W. Fields, and D. Alt. 1982. Land-based evidence for Tertiary climatic variations: Northern Rockies. *Geology* 10:413–417.

Tosk, T. 1988. Foraminifers in the fossil record: implications for an ecological zonation model. *Origins* 15 (1):8–18.

Vardiman, L. 1996. *Sea-Floor Sediments and the Age of the Earth*. Institute for Creation Research, El Cajon, CA.

Vardiman, L., A.A. Snelling, and E.F. Chaffin (editors). 2000. *Radioisotopes and the age of the Earth: a young-Earth creationist research initiative*. Institute for Creation Research and Creation Research Society, El Cajon, CA and Chino Valley, AZ.

———. 2005. *Radioisotopes and the age of the Earth: results of a young-Earth creationist research initiative*. Institute for Creation Research and Creation Research Society, El Cajon, CA and Chino Valley, AZ.

Walker, T. 1994. A Biblical geological model. In Walsh, R.E. (editor), *Proceedings of the Third International Conference on Creationism* (technical symposium sessions), pp. 581–592. Creation Science Fellowship, Pittsburgh, PA.

———. 1996a. The basement rocks of the Brisbane area, Australia: where to they fit in the creation model? *TJ* 10(2):241–257.

———. 1996b. The Great Artesian Basins, Australia. *TJ* 10(3):379–390.

Whitcomb, J.C., Jr. and H.M. Morris, 1961. *The Genesis Flood*. Baker Book House, Grand Rapids, MI.

Wieland, C. 1995. Sensational Australian tree…like 'finding a live dinosaur.' *Creation Ex Nihilo* 17(2):13.

Willis, B. 1902. Stratigraphy and structure, Lewis and Livingston Ranges, Montana. *GSA Bulletin* 13:305–352.

Wise, D.U. 2000. Laramide structures in basement and cover of the Beartooth uplift near Red Lodge, Montana. *AAPG Bulletin* 84(3):360–375.

Woodmorappe, J. 1996. *Noah's Ark: a feasibility study*. Institute for Creation Research, El Cajon, CA.

———. 1999a. *The mythology of modern dating methods*. Institute for Creation Research, El Cajon, CA.

———. 1999b. A diluviological treatise on the stratigraphic separation of fossils. In *Studies in Flood Geology, second edition*, pp. 23–75. Institute for Creation Research, El Cajon, CA.

———. 1999c. An anthology of matters significant to creationist and diluviology: report 2. In *Studies in Flood Geology, second edition*, pp. 79–101. Institute for Creation Research, El Cajon, CA.

———. 1999d. The essential nonexistence of the evolutionary-uniformitarian geological column: a quantitative assessment. In *Studies in Flood Geology, second edition*, pp. 105–130. Institute for Creation Research, El Cajon, CA.

———. 1999e. An anthology of matters significant to creationism and diluviology: report 1. In *Studies in Flood Geology, second edition*, pp. 133–144. Institute for Creation Research, El Cajon, CA.

———. 1999f. The cephalopods in the creation and the universal Deluge. In *Studies in Flood Geology*, second edition, pp. 179–197. Institute for Creation Research, El Cajon, CA.

Author

Michael Oard earned his M.S. degree in Atmospheric Science from the University of Washington. He retired after 30 years as a professional meteorologist with the National Weather Service. He has published numerous articles in creationist periodicals and has authored or coauthored nine creationist books.

Forum Discussion

1. If the author would follow the distinction between the empirical rock record and the conceptual geologic column (Chapter 8), his discomfort with abandoning the column might be lessened. Empirical order found in the rock record does not demand the column. (Reed, Klevberg, Froede)

I find no discomfort in abandoning the column. I advocate using Walker's biblical geological model to classify sedimentary rocks within stages and phases of the Flood, and let the periods of the geological column fall where they may. I agree that the geologic column is conceptual, but it seems like the empirical rock record, which does not demand the geological column, nevertheless generally lines up with the conceptual column, with many exceptions of course.

2. If the "Cenozoic" can be anything from early Flood to post-Flood, the geologic column as a rigid system of correlative global synchronous time units is invalid. (Reed, Klevberg, Froede)

I agree that the Cenozoic is invalid as a global synchronous unit. There seems to be regional systematic relationships, however. The "Cenozoic" in the Rocky Mountains and High Plains seems to be generally early Flood, because of great Recessional Stage Flood erosion that has carved down to the level of these rocks. The "Cenozoic" of the continental margins likely is Recessional Stage deposits, while some of the "Cenozoic" of the deep ocean basins and submarine fans is likely post-Flood.

3. While Oard asserts a fossil order in the empirical rock record, his factual basis is questionable. In Montana and adjacent southern Canada, at least one formation has apparently been "split" in two to preserve the column due to laterally proximate Triceratops and mammal fossils, and scores of physically baseless overthrusts invoked to accommodate out-of-order marine fossils. Thus, the "data" provided by evolutionists are likely to be misleading unless field checked by diluvialists. (Reed, Klevberg, Froede)

I mentioned overthrusts as likely examples of out of order strata, but creationists need a large research project to sort out "overthrusts." I also suspect this selective labeling of Cretaceous and early Cenozoic strata, such as in Canada, has been done with mammals and dinosaurs on the High Plains, but it would be difficult to prove. As far as an empirical basis for fossil order, I mentioned the Bighorn Basin in which practically all the periods are found in correct order from the Precambrian granites to the Cenozoic valley fills in the center of the basin. There is also the Grand Staircase in the southwest United States that preserves the "Paleozoic and Mesozoic" in correct order.

4. You have indicated that lithology is a means of correlating local columns and suggest that problems start when index fossils are used apart from lithology. How far can lithology take us? (Tyler)

Good question, especially when there may not be a correlation between index fossils and particular lithologies. But, lithologies may work for local columns, since in southwest Montana and vicinity the "Paleozoic and Mesozoic" strata seem similar in the various mountain ranges. The "Precambrian" Belt Supergroup has a fairly distinctive lithology, texture, and other features. However, lithology may fail us in the Cretaceous and Cenozoic due to a large variety of rocks and rapid facies changes. I would rather use lithology than index fossils, based on evolution and fossil succession, to set up a local column in my area.

5. Is it fair to read your chapter as saying that that the broad Precambrian/Paleozoic/ Mesozoic divisions of the geologic column are meaningfully correlated in terms of exposed vertical sequences of strata? If so, how can this be reconciled with your comments about the geological column being "mostly a horizontal fossil scheme"? (Tyler)

I was thinking of the global column when I said that the geological column was mostly a horizontal scheme. But, even in a regional or local rock column, much of the fitting together into a vertical scheme is from lateral sections.

6. Regarding Walker's model, you refer to tracks being one of his "defining criteria for the Inundatory Stage." From a biblical perspective, there are two opportunities for tracks of air-breathers to form: early in the Flood and during post-Flood recolonization. How then can tracks be a "defining criteria" for early-stage Flood sedimentation? (Tyler)

I certain agree that tracks are from either early Flood rocks or post-Flood sediments; all air-breathing land animals are dead (except those on the ark) and cannot make tracks between Day 150 and Day 371. To determine the time frame, a creationist must take into account other variables. For instance, hundreds to thousands of meters of strata have been removed from above dinosaur tracks in the Rocky Mountains and High Plains of the United States. Such erosion is typical of the Recessional Stage of the Flood.

7. Your conclusion that "Flood sedimentation is highly nonlinear with most sediment deposited in the Inundatory Stage" strongly suggests that the early stages of the Flood were considerably more violent than many have previously thought. Do you agree that this violence undermines attempts to account for fossil successions based on mechanisms of ecological zonation and differential mobility? (Tyler)

No, I do not agree that the early Flood was so violent everywhere that there were no refuges, high lands not yet flooded, or tectonic uplifts to expose areas, etc. I also believe that general burial by ecological zonation is a reasonable concept. As far as differential mobility, it is interesting that after the Indian Ocean tsunami, many animals were found on top of hills. There were likely many other variables that would contribute to fossil distribution.

eight

Interpreting the Rock Record without the Uniformitarian Geologic Column

John K. Reed
Peter Klevberg
Carl R. Froede Jr.

Introduction

What is the role of the global uniformitarian stratigraphic column in Flood geology? This book reflects a wide range of opinion among creationists and there is a wealth of additional discussion in the literature.[1] Perhaps the "frontier" nature of diluvial geology, like many other innovations, has forced this surfeit of ideas. But a lack of clarity at the axiomatic levels of natural history and stratigraphy (Klevberg 1999; 2000a; Reed, 2001) also plays a role. Therefore, while we hope to persuade creationists to reject the column, it is more important that we focus attention on underlying questions, because we are concerned that many creationists are content to address this issue on the 'home field' of the uniformitarians; explicitly rejecting their conclusions, but unconsciously accepting many of their assumptions about time, history, and reality.

Over the past decades, creationists have approached the column like contractors disagreeing about the extent of damage to a house; some suggest renovation, while others recommend reconstruction. We agree with Whitcomb and Morris (1961, pp. 115, 119–120), who cast an early vote for the latter:

> It is the conviction of the writers … that a true historical geology will never be formulated un-

[1] In addition to the summaries in this book, we recommend the following representative works. Those questioning the use of the column: Burdick (1979), Froede (1995; 1998), Jackson (1980), Morris (1996), Reed (2001), Walker (1994), Woodmorappe (1996; 1999a; 1999b). Those affirming the reality and applicability of the column: Garner (1996a; 1996b), Garton (1996), Northrup (1977; 1979; 1986; 1990a; 1990b), Robinson (1996; 1997), Snelling (1996), Snelling et al. (1996).

til the Genesis Flood … is granted its rightful and vital place … *It becomes very important, therefore, for Christians to re-study and re-think the great mass of geologic and paleontologic data, with two main purposes in view.* The first aim should be to examine carefully the currently accepted scheme of historical geology and its guiding principles, in order to determine clearly wherein and to what degree it is at variance with the biblical record of creation and the Flood …

The second aim … will be to develop, if possible, a new scheme of historical geology, which would not only be true to the biblical revelations that are pertinent to it but also would serve as a better basis of correlation for the available scientific data than does the present one … *There is an immense amount of data available that must be restudied and re-evaluated, enough to require the attention of many experts for a very long period of time* (emphasis added).

Following the sketicism of Price (1923), they clearly envisioned a reappraisal of stratigraphy down to the level of the data and a subsequent rebuilding of many aspects of geology. We believe that uprooting uniformitarian foundations requires no less. Some creationists have followed their advice, while others have opted for a less drastic path, but it is clear that there is no current consensus on even the scope of diluvial geology; much less its general content and methods (Table I).

Author	Approach to Column
Whitcomb and Morris, 1961	Column questionable; Genesis Flood caused majority of rock record
Clark, 1968	Column indisputable; young earth; geologic eras represent multiple catastrophes alluded to in Bible
Northrup, 1977; 1986; 1990	Column is real; multiple catastrophes in biblical history correspond to major divisions of column
Brown, 1980; 2001	Rock record a result of tectonics and liquefaction
Woodmorappe, 1981; 1983; 1999a; 1999b	Replace column with TAB (Tectonically-Associated Biological Provinces) model
Austin et al., 1994; Baumgardner, 2002	Use relative dating from column to discriminate ages of sea floor and continents for plate motion
Walker, 1994; Froede, 1995	Biblical columns prior to empirical investigation
Austin and Wise, 1994	Set discontinuity criteria for Flood boundary and applied to Grand Canyon
Snelling et al., 1996	Use column with young-earth chronology
Reed et al., 1996	Geologic energy, not time, is stratigraphic key
Gentet, 2000	Creation, Curse, Catastrophe model accepts column; eras are paleoecological zones

Table I. Creationist Approaches to the Column

Some prefer that diluvialists simply wallpaper over the absolute timescale, preserving the rest of the column for reasons of efficiency and utility. Robinson (1997, p. 14) stated:

> If this view [rejecting the column] is correct, a geologist who believes in the Flood is obligated, as he seeks to make sense of the earth's strata, to retread the path taken by geology and start again, from where a wrong turn was taken in the 18th century…. (brackets added).

Though his argument may seem pragmatic, it invokes the chronological fallacy, confusing truth with tradition. If geologists have built error into the column, then its cumulative complexity makes it no less false. Should Copernicus have turned back because of all those centuries of sophisticated Ptolemaic analysis?

Nonetheless, this argument strikes a chord with creationists who are few in number and lack the means of their peers in academia, government, or industry. The column seems to offer leverage for their limited resources; a shortcut to highly-developed models that can compete with those of uniformitarian geology. Their goal is commendable and their concern is valid: any substantial reconstruction of stratigraphy will demand great effort and may offer little immediate return. For that reason alone, we stand ready to be convinced that renovation is the best option. If creationists can hold to truth with no more than minor adjustments to the timescale of the column, then let's renovate! Unfortunately, the logical necessity of a more drastic approach seems clear. If truth demands the difficult path, then it is past time to begin clearing away centuries of error. Either way, truth—not utility—must guide our decision.

Since truth can be complicated, it is always helpful to revisit the foundations and regain perspective. Simply put, rebuilding natural history within the biblical worldview requires replacing evolutionary and uniformitarian history with creation, Flood, and a young Earth. We contend that this task demands more than simply compressing the timescale; it requires abandoning the global geologic column. Our case can be summarized as follows:

(1) The conceptual foundations of the column must be addressed, especially its extrascientific presuppositions.
(2) Those foundations present lethal logical incompatibilities with the biblical worldview.
(3) Logical errors outweigh empirical and utilitarian arguments.
(4) Empirical data can be better interpreted without recourse to the column.

Our argument extends beyond a negative critique. The beginnings of a sound diluvial approach can be found in several positive proposals. These include: (1) resolving confusion over the nature of the column by clear terminology; (2) laying a firm foundation explicitly within the biblical worldview; and (3) resolving the fate of the column from the standpoint of the resulting biblical natural history.

First Problem: The Fallacy of the Empirical Modular Column

Creationists often seem to be talking at cross-purposes when they discuss the column. One group points to an outcrop or well log and says, "Here it is!" Others speak of philosophical abstraction. At the heart of this confusion is the belief by some that the column is: (1) empirical and (2) divisible. They contend that it is composed of observable components readily apparent in the field. We are convinced that the column is neither empirical nor divisible in the manner advocated, and suggest a more precise vocabulary to clear away this fog.

First, is the column divisible—comprised of modular components that can be rearranged or replaced without detriment to the whole? Some think so.

> We recognize the general validity (but are cautious about the details) of the lithostratigraphic and biostratigraphic columns, question the assumptions in the chronostratigraphic column, and reject entirely the absolute geological timescale (Snelling et al., 1996, p. 334).

Picturing the geologic column as a synthesis of constituent lithostratigraphic, biostratigraphic, chronostratigraphic, and geochronologic components may be convenient for creationists who believe that an instant diluvial column can be achieved by replacing the faulty geochronologic module which outputs billions of years, with one that measures them in thousands. However, their proposal ignores at least two problems. First, the column cannot be both empirical and synthetic at the same time. A synthesis implies an overarching plan guiding the assembly of the various parts. Even if the parts were empirical, such a framework most definitely is not. Furthermore if the framework transcends the data (dictating their ordering), then altering the synthesis will require modifying the framework, not just exchanging parts. You would not give a soccer team a smaller ball and expect that alone to start them playing baseball!

Second, those who built and maintain the column do not comprehend or apply it in this modular fashion. They may talk about *local* lithostratigraphic and biostratigraphic columns, but they uniformly assert a global chronostratigraphy as the glue holding the pieces together. Amazingly, each of their local constituent columns always ends up displaying the same history—not because of a miraculous empirical similarity, but because the master design demands it. The global column determines the structure of its constituents, not the other way around. Uniformitarian stratigraphy explicitly upholds uniformitarian natural history and like an arch, we cannot remove particular components without causing the entire structure to crumble (Figure 1).

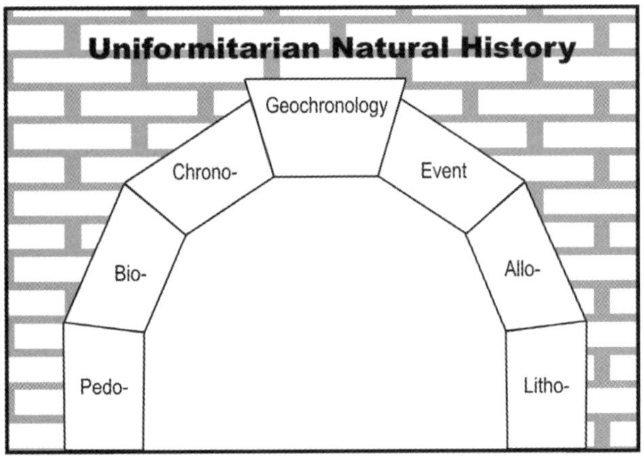

Figure 1. Uniformitarian stratigraphy must bear the weight of uniformitarian natural history. Once a particular "stone" is emplaced in the system, it cannot be removed without precipitating the failure of both the arch and the wall it upholds.

As an aside, note that the so-called components of the column are simply conclusions reached by separate stratigraphic *methods*: lithostratigraphic columns are derived from lithostratigraphy, biostratigraphic columns from biostratigraphy, etc. Each of these methods contains built-in axioms of naturalism. For example, biostratigraphy is built around the assumption of evolutionary progression. Stratigraphers "see" the progression because they "know" that evolution has occurred. A related problem is epistemological. Linking facets of the column to methods implies that truth is based on methodology—a position entirely congruent with the positivism of naturalism, but not with the biblical worldview. Advocates of naturalism believe that truth is attained by filtering data through correct methods, but creationists cannot so easily ignore the hidden extrascientific baggage. Each worldview sees history through its own lens. We cannot accept naturalism's casual merger of data and conclusions or of science and history.

If the column is not modular, then is it empirical? Some say yes.

> The 'geological column'—that is, the raw data of the earth's rocks interpreted by geological investigation into some sort of order… (Robinson, 1997, p. 20).

However, the author refutes his own assertion by using the word "interpreted." Interpretation is not observation. It is the application of a framework of presuppositions onto limited and selected data. As Clark (1994, pp. 21–22) noted, "History requires philosophy… It can be ignored, but it cannot be avoided." Equating the column to the rocks in the field is like equating a car to its steel, ignoring form and function. Likewise, the problems that creationists face are far deeper than simply the quantity of time. Like memory metal, the column is prestressed by evolution, deep time, and uniformitarianism. No matter how it is twisted into a diluvial model, it will always revert to its original form.

In reality, the column is not so much empirical as it is trans-empirical: it functions like a Platonic ideal—a stratigraphic, sedimentologic, and paleontologic type section of the crust absent erosion and nondeposition. It includes a framework not found in the physical description of the rocks themselves: a definite, linear historical sequence of absolute ages, paleoenvironments, tectonic motions, geomagnetic fluctuations, etc. Since *any* historical proposition contains non-scientific assumptions (Clark, 1994), history is by definition interpretive in a manner that science is not. Though uniformitarians try to cloak their history in the certainty of science; creationist history must rely instead on an explicit link to revelation.

If the column were an empirical construct, would it not be constantly changing as new data are acquired? Because it has been the dominant paradigm for decades, many seem to forget that inherent historical uncertainty cries out for alternate explanations. The fundamental inflexibility of the uniformitarian column smacks of dogma, not unbiased empirical investigation, especially in the face of contrary empirical (not to mention historical, theological, and philosophical) evidence.

Finally, since the column's neutrality is obviated by its close association with naturalism, creationists can apply it only after they identify and resolve discrepancies between Christianity and naturalism. Since logical validity depends on this step, it must take precedence over gathering data. After all, an elegant model that violates axioms of contradiction and causality is no less invalid than a simple model that does the same.

If the present debate is stuck on whether or not the column is empirical and divisible, then how can creationists advance diluvial geology? We believe part of the answer lies in vocabulary. Words, especially technical terms, require careful definition. Part of the current confusion can be blamed on imprecise terminology. Those who see the column as an empirical synthesis see the call to reject it as a demand to reject the rock record, the ability to correlate rock units, the significance of fossils, geophysical signatures... indeed, geology itself![2] They seem frustrated that others cannot understand that they *see* the column in the field. On the other hand, those concerned with the conceptual baggage of the column fear a morass of compromise with uniformitarianism and evolution. These different points of view explain why creationists often seem to be talking about two different ideas.

The debate can and should be advanced by forcing clarity at this point. All creationists should be able to agree that distinguishing between data and interpretation is a worthy goal. Therefore, we have offered terms that reflect that distinction: **rock record** and **geologic column** (Froede and Reed, 1999). **"Rock record**" is a descriptive term for those portions of the earth's crust open to human examination and **"geologic column**" is the interpretation of the rock record by means of assumptions about deep time, evolution, and the continuity of depositional processes operating in modern environments. Due to the many differences between uniformitarian and diluvial stratigraphy, and in the interest of minimizing confusion, we suggest that those creationists who want an alternative column coin a new term for their constructs other than "geologic column."

This change in vocabulary rests on the rock record being *perceptual* and the column, *conceptual*. When geologists talk about seeing the column in the field, they are obviously referring to the rock record. Unfortunately, most geologists are so familiar with

[2] If nothing else, our previous work—theoretical and practical—refutes that misperception. See Froede (1995; 1998; 2005a; 2005b), Froede and Reed (1999), Klevberg (1998; 1999:2000a; 2000b), Klevberg and Oard (1998; 2005), Reed (2000; 2001; 2002a; 2003; 2004; 2005), and Reed and Froede (1997; 2000; 2003).

Figure 2. The Grand Canyon (left) shows the perceptual rock record of the Colorado Plateau in cross section. It can be defined by the geometry and composition of the rock units. The first interpretive step is to classify the rocks into rock units, based on their properties. The second is more of an interpretive leap—ordering those units within a historical template based on assumptions of evolution, uniformitarianism, and deep time. The conceptual column (far right) is quite distinct from the perceptual rocks; it is an interpretive overlay of their history. Column modified from Beus and Morales (1990).

the conceptual overlay that little thought is given to the differences between what they see with their eyes and what they see in their mind. Those differences are more obvious when seen side by side (Figure 2). Although this distinction is not often discussed, it is crucial. We are convinced that until it is widely accepted, creationists will continue to argue past each other.

To be sure, there is overlap between the rocks and the column, but that is only to be expected. After all, the column interprets the rock record. Likewise, any future diluvial construct will incorporate the rock record, though it might be otherwise distinct from the uniformitarian column (Figure 3). While it is naïve to ignore the features of the rock record, it is more naïve to not see the column's conceptual components. Like Aaron's explanation of the golden calf of the exodus, many seem to think that geologists of yore simply threw the rocks and fossils into the fire of observation and out popped the column! But like Moses, we must be more skeptical. An outmoded positivist view of natural history must be replaced with a more realistic one that recognizes the relative roles of revelation, theology, philosophy, history, and science.[3]

We certainly do not dismiss the field research of the past two centuries. In fact, we see similarities between creationist and non-creationist stratigraphy; if nothing else they are investigating the same phenomena with many similar methods. We hold no brief against uniformitarian observations, only against their template of natural history. Other points of commonality exist, too. Just the willingness to use the term, "rock record" implies that both sides accept the assumption that historical information can be

[3] For additional information on the relative roles of theology, philosophy and science, we recommend: Adler (1965), Hasker (1983), Middleman and Wilder-Smith (1980), Pearcey and Johnson (2004), Plantinga (1990; 1997a; 1997b), Reed et al. (2004), Schaeffer (1968; 1976), Titus et al. (1979), Wilder-Smith (1988).

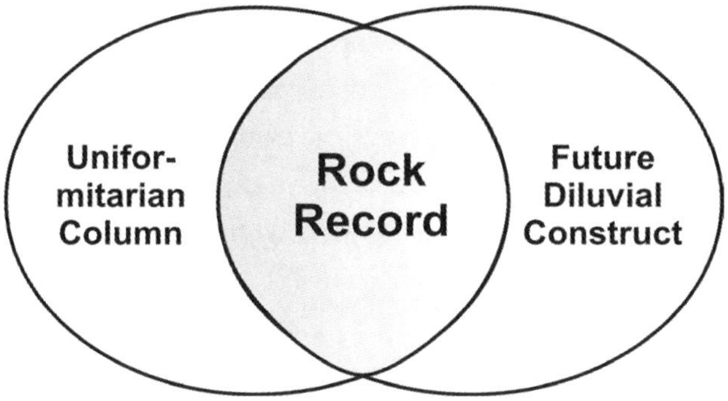

Figure 3. There is a logical overlap between the column and the rock record because the column is a conceptual explanation of the rocks. The two should not be confused; a similar overlap would exist between the rock record and a diluvial construct.

derived from examination of the rocks. However, these points of commonality should not blind anyone to myriad contrary assumptions between the worldviews embedded in the column—assumptions that create logical inconsistencies between the column and diluvial geology.

Second Problem: Logical Inconsistencies

It is generally agreed that the law of contradiction is a cornerstone of truth in human thought. Inconsistencies between the Bible and the tenets of evolutionism and uniformitarianism are unavoidable. Thus, our position can be summarized as follows:

> *If the column rests on the presuppositions of evolution, deep time, and uniformitarianism—all inimical to creationism—then it has no meaningful role in creationist stratigraphy.*

Clearly, logical contradictions between presuppositions carry greater evidentiary weight than empirical data, if for no other reason than that data must be interpreted by logical categories. Those advocating the use of the column tacitly admit our position whenever they insist that the column is independent of evolution and uniformitarianism. Therefore, we need only correct that mistaken impression to gain their agreement with our thesis.

Those points need to be carefully addressed since some creationists believe that:

> ...the lithostratigraphic and biostratigraphic columns can be constructed... without using "evolutionary" and "uniformitarian" presuppositions (Tyler and Garner, 2000, p. 60).

Such columns might be possible as thought experiments, but we need to deal with the column as it *is*, not as we might wish it to be. And it *is* a unified construct predicated on these assumptions. The inability to see those connections may well stem from the initial failure to discriminate between the rock record (observable phenomena) and the column (historical template of naturalism). While it is clear that rocks and fossils are the building blocks of the column, they are arranged according to the blueprint supplied by naturalism's "contractors"—evolution and uniformitarianism.

Even a cursory examination of evolutionism and uniformitarian geology reveals glaring presuppositional differences with diluvial geology—differences that run much deeper than the absolute age of the crust. These come into focus most easily when viewed through the lens of parent worldviews. Links between the column and uniformitarianism/evolutionary progression are well documented by both proponents (Bliss et al., 1980; Clough and Fredericks, 1978; Gish, 1985; 1995; Wilder-Smith, 1993) and opponents (Berry, 1968; Harbaugh, 1974; Jablonski et al., 1996; Gould, 1997) of diluvial geology.

In the following section we assume that Earth's history included a global Flood. Informed inference would lead us to believe that events of that catastrophe produced the greatest levels of geologic energy (i.e., erosion, sediment transport, deposition, new sediment production, volcanism, tectonism, turbidites, extraterrestrial impacts, sea level changes) experienced by the planet since its creation, and resulted in the formation, preservation, and subsequent alteration of most of the sedimentary (and many of the igneous and metamorphic) rocks found in the crust (Reed, et al., 1996). Although some creationists have debated this point (Gentet, 2000; Setterfield and Setterfield), it seems an inarguable biblical inference.

Time, Uniformitarianism, and the Column

A crucial feature of the column is its emphasis on time. Whether the duration is measured in thousands or billions of years, the fact remains that the goal of correlation within the column is chronological. As Lemon (1991, pp. vii) stated:

> Stratigraphy is concerned with stratified rocks as the repository of a record of past events. It deals with sediments, sedimentary rocks, sedimentary environments, and fossils, and with the *ways they can be described in terms of passing time* (emphasis added).

Therefore, to assert that lithostratigraphic and biostratigraphic columns exist independent of a chronostratigraphic column ignores the fundamental fact that even if the column could be compressed to fit a young Earth history, it would still be a single vertical model correlating specific strata across the globe based on their being time equivalents. For example, layer 'X' in the Paris Basin would necessarily correspond to layer 'X' in the Gulf of Mexico, North Africa, India, etc. purely on the basis of their *chronological* identity.

Is time even the proper primary focus of diluvial stratigraphy? Much more than just its extent is at issue. Quantitative differences between billions and thousands of years demand qualitative differences in other areas. For example, diluvialists and uniformitarians diverge on the meaning of missing section. Diluvialists conceive of unconformities primarily as paleohydraulic phenomena; uniformitarians see gaps of millions of years. Also, we cannot ignore the elevated uncertainty demanded by the absence of much of the record (Ager, 1993; Woodmorappe, 1999a). Diluvialists—confident in their biblical foundation—do not seem to grasp the extent to which the paucity of physical evidence poisons the position of uniformitarians who insist on science as the hallmark of truth. When confronted with the absence of most of their record, uniformitarians use the sheer quantity of data as a red herring to disguise the absence of *expected* information, since the ratio between observed rocks and the total rock record of Earth's history is astonishingly low. Catastrophism continues to exaggerate that difference. For example, Austin's (2004) discovery of a massive nautiloid extinction in the Redwall Limestone strongly implies that part of the formation was deposited in hours, leaving yet another vacuum of millions of lost years. As creationists continue to demonstrate the rapid deposition of particular pieces of the rock record, uniformitarians will have to either give up deep time or face an almost nonexistent record.[4] The assertion that correlation allows the empirical development of an ideal section (which in turn obviates the need for local data) is becoming threadworn. The analogy of stratigraphy to a jigsaw puzzle—given enough pieces, interpolation can deduce the picture—will have to be refined. The truth in that analogy is that uniformitarian stratigraphers "see the picture on the cover" before they ever begin to assemble the pieces.

Missing section creates an additional problem for uniformitarians. Clearly there can be no knowledge of environments that might have existed during periods of erosion or non-deposition. Logically then, they cannot be certain that the "missing section" would not show conditions atypical of the "present," invalidating their cornerstone of uniformitarianism. Such problems are not addressed; uniformitarians simply default to an indemonstrable confidence that the present is the key to the past and cheerfully argue from a lack of evidence.

[4] Reed (2002b; 2004a) noted the consequences of evidence of catastrophic erosion and deposition decreasing the historical record of time.

In addition to problems presented by missing section, creationists need to consider the difficulties that diachronous sedimentation presents to an attempt to salvage the column in the context of a one-year Flood. Obviously time-transgressive deposition within a stage of 20,000,000 years will be less significant than within a period of days. This is true of any process that varies laterally with time. If total time is large, then the resulting impact in the vertical chronology of such a process will be small; if time is limited, then the impact increases.

Another crucial difference between the diluvial and uniformitarian view of history is the distribution of the empirical record. The column assumes a gross similarity of depositional rates over time. Both the rocks and the gaps are assumed to be evenly distributed across history's timeline (Figure 4). This position has a certain internal logic. Obviously, a representative record over time requires a statistically-significant sample, and the distribution of the sample population affects that significance. Contrary to the evenly-distributed data of uniformitarianism, the biblical record presents the distribution of the rock record as a sharp spike—representing only 0.017% of history if one assumes a 6,000-year total. Think of natural history as a movie. If future archivists unearthed scraps of film, the uniformitarians among them might believe that they had recovered one frame in ten evenly distributed throughout the reel, and thus think that they could reconstruct the story. However, their catastrophist colleagues might recognize that those fragments were a nearly-continuous few seconds somewhere in the middle (with perhaps a few frames on either side), and conclude that forensic study would be virtually useless in determining anything other than that particular scene.

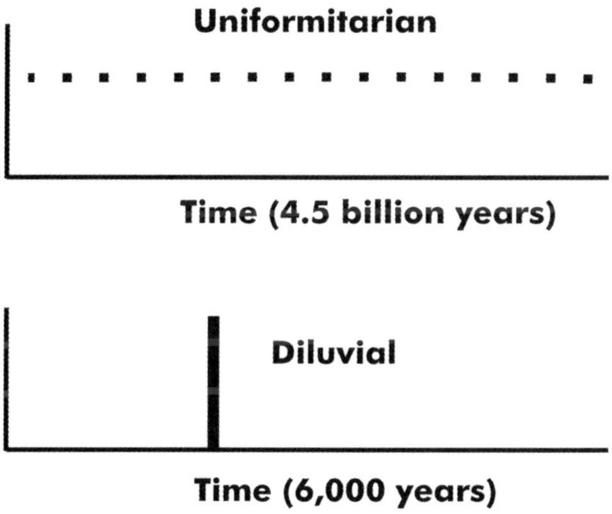

Figure 4. There is a stark contrast with regard to data distribution through time between the uniformitarian and diluvial paradigms. Though uniformitarians admit that most of the rock record is missing, they believe that there exists a representative sample through time, from which they build the column. However, diluvialists cannot relate the data and time in the same way, since most of the rocks form a data "spike" in the historical timeline.

This issue comes into play when creationists claim that the column can be compressed into the timescale of a young Earth. The different styles of data distribution alone suggest such compression is unworkable. There is no way to perform a one-to-one correlation of 'X' million years on the column to 'X' thousand years for a creationist analogue. In the former, there is uniformity in the processes producing the rock record; in the latter there is not. Similarity between the column and any Flood model would then be like equating two locations that receive fifty inches of rain—one in a regular annual succession of gentle showers and the other in a hurricane. The total amount of water might be the same, but its effects would be quite distinct.

Thus, both the amount of missing data and its style of distribution on Earth's timeline demonstrate the necessity of a biblical foundation for any successful natural history. Uniformitarians are pushing a boulder uphill. Their empirical approach to history demands data, and the absence of so much of it degrades their certainty. In the face of so much missing section, they must cling to a style of distribution that offers statistical hope. Diluvial geology is at odds with that scheme: (1) in its reliance on truth beyond empiricism and (2) in its insistence that the rock record is a data "spike" on the timeline. This incompatibility between the two paradigms is itself a powerful argument against the use of the column in diluvial research.

If the mass of the rock record was deposited rapidly and at wildly varying rates in different locations, then how can strata be considered globally correlative in a time-stratigraphic sense? We believe that the progression of geologic energy levels, reflected in constituent sections of the rock record, are a potential key to better defining events of the Flood. For example, the breakup of the fountains of the great deep might provide widespread geologic markers. Alternatively, there might also have been progressive changes in water chemistry during the Flood that could provide a means of correlation (Reed, 2002a).

While local relative timing can be inferred from superposition (with caution), crosscutting relationships, etc., extrapolation to a global chronology has not yet been demonstrated—merely assumed. Inferences from empirical data should be restricted to chemical/physical conditions (e.g., minimum current velocity) that can be tied (with varying levels of confidence) to observed features. Diluvialists should agree that even those conclusions face the uncertainties inherent in a unique event exhibiting a supernatural component. Thus, from a creationist viewpoint, the quantity and quality of history that can be derived from the rocks is currently unknown and should be expected to be more limited than that advocated by uniformitarians.

A related problem is the disparate perception by diluvialists and uniformitarians of rocks as the products of events rather than processes (Reed and Froede, 1997). The column is predicated on the uniformitarian assumption of depositional processes, operating in a similar manner to modern settings over long time periods (c.f., Walther's Law). On the other hand, the Bible posits the catastrophic emplacement of rocks, predominantly in a unique Flood. Picture the daily processes of waves and tides operating on a barrier island. Sediment is eroded, transported, and deposited in a series of well-defined environments and an idealized vertical succession of sediment types can be extrapolated. However, that "uniform" work of thousands of days can be erased or overwhelmed by a single large storm (e.g., Froede, 2005b; in press). If the majority of strata were laid down catastrophically, then all of the carefully reconstructed "modern" environments of deep time are a sham and diluvialists must look for markedly different explanations.

Finally, naturalism demands historical continuity *within* nature. It assumes that continuity is provided historically by the continuous operation of processes that can be scientifically described or extrapolated via uniformitarianism. Continuity is manifested biologically by the inexorable march of irreversible evolution. In contrast, the biblical model of history is one of physical discontinuity resulting from supernatural intervention, the bane of naturalism. For creationists, the possibility of history rests on the continuity derived from an eternal, omniscient God. So yet again, the two systems are incompatible.

In summary, the severely limited duration and the catastrophic nature of the biblical Flood seem to preclude the possibility of an ideal global vertical succession of laterally isochronous strata. It is not simply the duration of the chronostratigraphic column that we must question—it is its very possibility. Even a "compressed" timescale does not guarantee laterally-correlative synchronous sediments. Correlation can only be determined directly and the work of Berthault (1997) cautions against a universal application of even Steno's assumptions. Furthermore, when the rocks are absent or unobservable, certainty decreases. That is not to say that we reject stratigraphy. Rather, we believe that interpreting the events of history depends on first rejecting the uniformitarian fixation of time, not simply its long duration. Instead, diluvial geologists need to pursue a more rigorous empirical investigation of rock units in all of their properties, including geometry, lithology, biotic content, etc. Only when that work is integrated without the presuppositions of deep time and uniformitarianism, will a diluvial stratigraphy be possible.

Evolution and the Column

The necessity of evolution to the column is demonstrated by an unbreakable chain of logic: from evolution to predicted fossil order to stratigraphy and finally to history (Lemon, 1991, pp. vii–viii):

> Stratigraphy is concerned with stratified rocks as the repository of a record of past events. It deals with sediments, sedimentary rocks, sedimentary environments, and fossils, and with the ways they can be described in terms of passing time. It provides the means of piecing together a calendar so that world geologic records can be brought together into a coherent whole.... Modern stratigraphy would be *impossible* without biostratigraphic control... (emphasis added).

Or, as Stephen Jay Gould (1997, pp. 157–158) explained:

> A chronometer of history has one, and only one, rigid requirement—something must be found that changes in a recognizable and irreversible way through time, so that each historical moment bears a distinctive signature. Geologists have long appreciated this principle in the abstract, but had not found a workable criterion... Moreover, rocks are simple physical objects formed by chemical laws and, as such, do not bear distinctive temporal signatures. Quartz is quartz—conjoined tetrahedra with a silicon ion in the center, surrounded by four oxygen ions, each shared with a neighboring tetrahedron. So it was in the beginning and is now, and ever shall be so long as nature's laws prevail. Cambrian quartz is not different from Pleistocene quartz.
>
> But life is complex enough to change through a series of unrepeated states. Today we attribute this irreversible sequence to the workings of evolution... (emphasis added).

This relationship is affirmed officially and unambiguously in the North American Stratigraphic Code (2005, p. 1557–1558):

> Biologic remains contained in, or forming, strata are uniquely important in stratigraphic practice. First, they provide the means of defining and recognizing material units based on fossil content.... Second, the irreversibility of organic evolution makes it possible to partition enclosing strata temporally.

It is almost a waste of paper to restate the obvious fact that this is the modern consensus. However, some creationists have stated that the column is independent of evolution because it is merely an empirical compilation of lithologies and faunal assemblages. Clearly that opinion is not shared by Lemon, Gould, the NASC, or modern stratigraphers too numerous to cite here. Still, our argument does not rest on an appeal to authority or majority opinion. Rather it rests on the internal logic of evolution providing the only reliable chronometric reference point for the column. You cannot decipher a code without a key. Gould was correct—establishing global temporal relationships throughout the rock record requires some kind of absolute reference. If biblical history is rejected, then evolution is the only possible unbreakable strand of history that can transform the rock record into an historical record. It bears repeating here that the failure to see the role of evolution in the column results from the mistaken idea of its empirical-intuitive basis, an error which in turn stems from a failure to distinguish between the rock record and the column.

Evolutionists believe that they have a clock against which they can mark history. However, their ability to correlate faunal assemblages is severely degraded with increasing scale. Diluvialists have noted before that a realistic appraisal of the actual data demonstrates how flimsy their approach is on the global scale. Evolutionists may claim otherwise, but correlation becomes more and more difficult with distance and suites of index fossils are chosen with the conclusion of evolution firmly in mind (Woodmorappe, 1983). Furthermore, creationists cannot accept the concept of dating by index fossils because such an admission implies evolutionary progression. Once the index fossil chronometer is set aside, either another basis for chronological comparison must be made or the column must be set aside.

Some put forward the specious argument that because Darwin's publication postdates the column, that there can be no causal connection between them. But timing of publication does not always follow the chronology of ideas. If nothing else, we know that decades elapsed between Darwin's own idea of evolution and its publication. Besides, emphasis on Darwin is a smokescreen. Evolution has a much longer history, flowering in the Enlightenment. For example, Diderot (1713–1784), the influential French thinker, accepted "Transformism," and Lamarck (1744–1829) heavily influenced Lyell (Barzun, 2000). Gould (1997) admitted that 19th century geology, dominated by the views of Hutton and Lyell, included biologic views of progress, common when Lyell published his *Principles* between 1830 and 1833. As Lubenow (2004, p. 150) noted:

> Evolution, as a philosophical concept, was well-known long before Darwin published. Desmond King-Hele, author of a biography of Erasmus Darwin...writes, "After 1794...statements on the principle of natural selection and evolution came fairly thick and fast."

The rock record and the concept of evolution are separable; the geologic column and evolution are not. Those who persist in seeing no difference between the rock record and the column will flounder with respect to combating uniformitarian/evolutionary natural history and will fail to grasp the vision of how great a revolution diluvial geology could ignite within natural history and stratigraphy... which leads to our third issue.

Third Problem: History, Science, and Stratigraphy

Earlier, we noted the confusion rising from the failure to distinguish between the column and the rock record. Underlying that error is another: the mistaken belief that empirical data alone can generate an interpretation of history. Correcting both requires the context of worldviews (Clark, 1994; Reed, 2001). For example, Marxists and Christians could observe exactly the same events, but their respective worldviews would probably force them to dramatically different interpretations. Natural history is no less driven by conflicting worldviews. Axioms of naturalism are built into the column, and axiomatic conflicts are rarely resolved within science. The biblical worldview delineates an approach to history far different from that of naturalism and its geologic column. This level of analysis reveals foundational flaws in naturalism (Pearcey and Johnson, 2004; Reed et al., 2004), which in turn further reduce confidence in the column.

Foundations of Natural History

One of the foremost problems in dealing with the column is an ongoing failure to set the debate firmly inside a framework of biblical presuppositions. Uniformitarian geology is sufficiently mature that its methods and assumptions are seldom questioned by its practitioners. Creationists do not have that luxury. We must recognize and evaluate both similarities and differences between biblical natural history and naturalism. First and foremost, diluvialists should recognize that natural history is *history,* undergirded and framed by theology and philosophy. For far too long, uniformitarians have blurred science and history and it is easy to slip into that mode of thinking. We address that relationship in Chapter 3, but suffice it to say here that in any biblical natural history, science must play in a secondary role, and must not exclude revelation, theology, history, etc. since natural history is a mixed question (Reed, 2001).

While naturalism considers science and natural history as synonymous (Cleland, 2001), creationists cannot. Historical reconstruction differs from experimental science, especially with regard to certainty (Table II). Furthermore, creationists do not recognize "prehistory," but see instead natural history as distinct from the biblical record not by predating it, but by seldom being the focus of the narrative. Most of natural history chronologically parallels human history, but focuses on a different source—the rocks. Creationists affirm the validity—indeed, the superiority—of the historical record in the Bible, while naturalists do not. Uniformitarian natural history, based on different presuppositions and methods, can also demonstrate profound effects on data selection and interpretation bias.

Therefore, creationists cannot succumb to the temptation to embrace the column to make the most of their limited resources. Instead, we must rigorously evaluate the foundations of uniformitarian natural history, replace those built on unbiblical axioms, and then ensure compatibility between those new axioms, the Bible, and methods and conclusions. Only then will stratigraphic research prove fruitful. We are no less eager than others to see the flowering of diluvial geology, but without firm roots, there will be no lasting fruit. As outlined above, that task will demand more than just the rejection of the geochronologic scale. Others have warned of the scope of the task (Clough and Fredericks, 1978; Whitcomb and Morris, 1961; Woodmorappe, 1981), but unfortunately, they have not been widely heeded.

Science	History
ongoing	unique
repeatable	unrepeatable
directly observable	not directly observable
primarily inductive	primarily deductive
relies principally on measurement/observation	relies principally on testimony/observer
failure to recognize limits leads to distortion and faulty conclusions	failure to recognize limits leads to distortion and faulty conclusions

Table II. Traits of Science and History

Boundary Disputes

In the time-weighted framework of the column, identifying time periods in the rock record is the essence of stratigraphy. Nineteenth-century stratigraphy was the time of great debates focusing on the placement of the boundaries within the time/rock record. Today, many secular geologists assume that boundaries are set and have shifted their emphasis to interpretation of ancient depositional environments (e.g., Berggren and Van Couvering, 1984; Brett and Baird, 1997; Donovan, 1989; Erwin, 1993; Hallam, 1992; Wilgus, Hastings, Kendall, Posamentier, Ross and Van Wagoner, 1988). In spite of this emphasis, they continue to operate within the framework of the column and its formal time/rock divisions. Unfortunately, many catastrophists, though advocates of a young Earth, have adopted this uniformitarian preoccupation and are proceeding to set their own "golden spikes" by correlating Flood boundaries to points on the column with the apparent expectation that these markers can be extrapolated globally.

If the column is incompatible with diluvial geology, then deriving boundaries at the "Cambrian/

Precambrian" contact (for example) is not only meaningless, but it implies a confidence in the uniformitarian construct that it does not deserve. A caution should be noted. While we believe that an attempt to save the time units of uniformitarian geology by correlating them to biblical events (i.e., Flood = Paleozoic) inevitably leads creationists into a morass, we recognize that diluvial geologists must apply the terminology of the column because at present it is the default language of stratigraphy. But there is a difference between referring to a *rock* unit as "Precambrian" and thinking that the same rock unit is also "Precambrian" in the sense of being a valid global *time-rock* unit.

Unfortunately, creationists have become caught up in pursuing "golden spikes" within the framework of the column, assuming global chronostratigraphic correlation between eons, eras, and stages. This exercise has been pursued in several ways. Some have simply advocated the use of a compressed column, and pursued the placement of pre-Flood/Flood (A/D) boundaries and Flood/post-Flood (D/P) boundaries (Garton, 1996; Garner, 1996a, 1996b; Holt, 1996; Robinson, 1996; 1997; Snelling, 1996; Tyler, 1997). They assume that Flood stratigraphic boundaries will be equivalent to the chronostratigraphic units of the column. In their mind, the main issue of diluvial stratigraphy is to determine whether the D/P boundary is, for example, the top of the Permian or the top of the Cretaceous. Others (Austin and Wise, 1994) have pursued a more thoughtful path—defining empirical discontinuity criteria to define the boundaries. However, they seem to imply that these boundaries will coincide with those established within the column. It is not clear if they intend to limit that determination spatially or think that such boundaries can be correlated worldwide based on the chronological divisions of the column. If global, should we expect correlative worldwide discontinuities or rubble zones? While we are willing to grant the difficulty of discussing the rock record without recourse to the nomenclature of the column, we fear that some proposals move beyond semantics.

All of these attempts have shared one important assumption—that the time-based stratigraphy of the column is compatible with the event-based stratigraphy of Genesis. Because the time available for geologic work is so compressed by the biblical record, efforts to understand the relationships between the rocks and time may be actually misdirecting workers away from more profitable investigations. A major shortcoming for creationists attempting to utilize the conceptual framework of time-based stratigraphy is their disposition to add multiple high-energy events to the Genesis Flood to explain the rock record. Although this does not necessarily violate the Bible, silent about many historical happenings, the rocks appear to demand events of such energy and magnitude (especially given the limited timeframe) that the uniqueness of the biblical Flood is called into question.

This approach is not new. A study of the historical debate that led to the discarding of diluvial geology in favor of modern uniformitarianism reveals a clear trend of compromise on the part of Christians that led in turn to the abdication of biblical authority in Earth history (Mortenson, 2004). In order to avoid the same mistakes, we believe that Christians must also avoid inferring an epistemological equity between natural history speculations and the Bible. This position inevitably leads toward the dynamic accumulation of "facts" supporting the column, making the sparse static accounts of the Bible appear worn and outdated. Historically, the increasing complexity of the column demanded other, less-catastrophic events in addition to the Genesis Flood in order to harmonize "science" and Scripture. In the end, Christians kept downgrading the Flood until it became little more than a caricature: an overflowing of the Euphrates River Valley (Sauer, 1996), the infilling of either the Mediterranean Sea (Morton, 1995) or Black Sea (McInnis, 1998; Ryan and Pit-

Figure 5. Exposed sidewall of the Thigpen Mine (kaolin) near Americus, Georgia, USA (10/2003). The rock record reflects (moving bottom to top) a transition from nearshore marginal marine to fluvial deposits. These have been interpreted as over 50 million years of uniform deposition (right) and as slow withdrawal of floodwater and transition to fluvial deposition (left). For additional detail, see Froede, 2005a, figures 2 and 3. Dashed lines in column to right indicate unconformities; diluvial stratigraphic column not to scale.

man, 1998), or even a tsunami associated with the eruption of Santorini (Myles, 1985). The essence of this misguided thought process was captured by Fields (1976, p. 184) where he lamented:

> There seems to be an assumption that if Christianity is to realize its full potential of impact on the scientific community, the message that no conflicts exist between the Bible and uniformitarian science must be heralded.

In our pursuit of ancient history, it might be well to remember the lessons of more recent ages.

Alternate Approaches

While the desire to incorporate the column into creationist stratigraphy may be present, the necessity is not. Creationists have proposed and applied other approaches. An early, exhaustive evaluation of the distribution of fossils by Woodmorappe (1983) led to the development of his "TAB" (tectonically-associated biological provinces) model. In 1994, Walker presented a stratigraphic model proposing a framework extracted from the Bible. He applied his model at locations in Australia (1996a; 1996b) and New Zealand (2001). Froede (1995) published a similar timescale, based on events described in Genesis and Job, and applied it to evaluate the feasibility of the column in the Gulf of Mexico basin (Froede, 1997; Froede and Reed, 1999), as well as to coastal plain sediments in southwest Georgia (Froede, 2005a) [Figure 5]. Reed, et al. (1996) advocated geologic energy as the integrating factor in creationist stratigraphy. The Walker, Froede, and Reed proposals are all deductive constructs, based on the Bible (Figure 6).

These theoretical alternatives have been complemented by field studies that have worked without applying the column. Akridge (1998; 2000) and Akridge and Froede (2000) described and interpreted strata found in the area around Lookout Mountain (NE Alabama-NW Georgia border) with reference to the Genesis Flood without relying on the column. Interpretation outside the column was also applied to gravel deposits found in Montana and sections of Canada (Klevberg, 1998; Klevberg and Oard, 1998; 2005; Oard and Klevberg, 1998) [Figure 7]. Reed (2000; 2002a; 2003; 2004b; 2005) described and in-

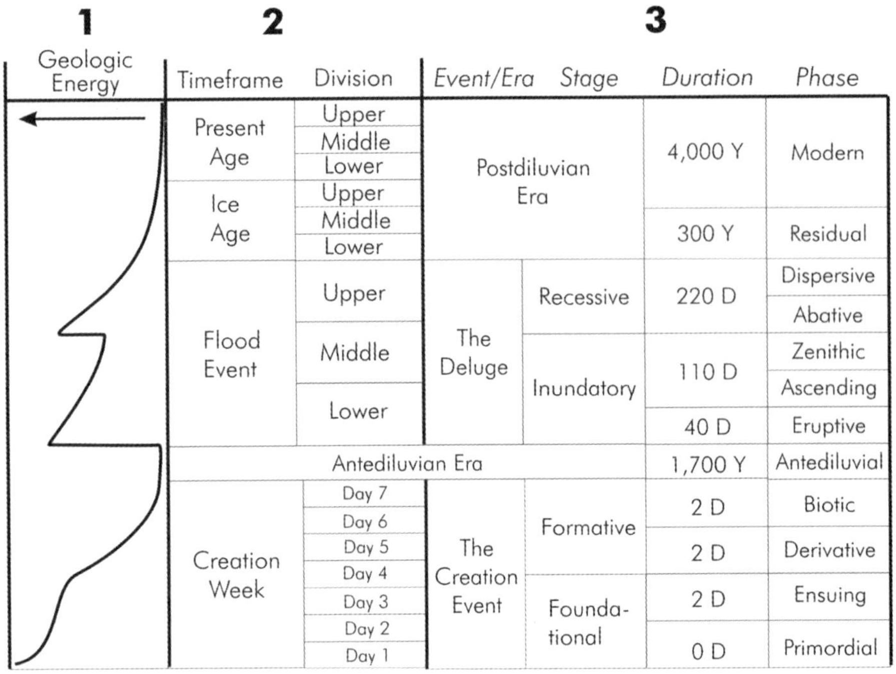

Figure 6. Deductive models of stratigraphy shown together for comparison. A = Reed et al. (1996); B = Froede (1995); and C = Walker (1994). Correlations may not be exact.

terpreted strata associated with the North American midcontinent with respect to tectonic and depositional energy levels rather than the column (Figure 8). These studies demonstrate that the column is not necessary to diluvial stratigraphy.

Conclusion

Though many creationists advocate the application of the column, they do so absent a thoroughgoing analysis of its conceptual aspects, especially its presuppositions. Drawing the crucial distinction between the perceptual rock record and the conceptual column brings

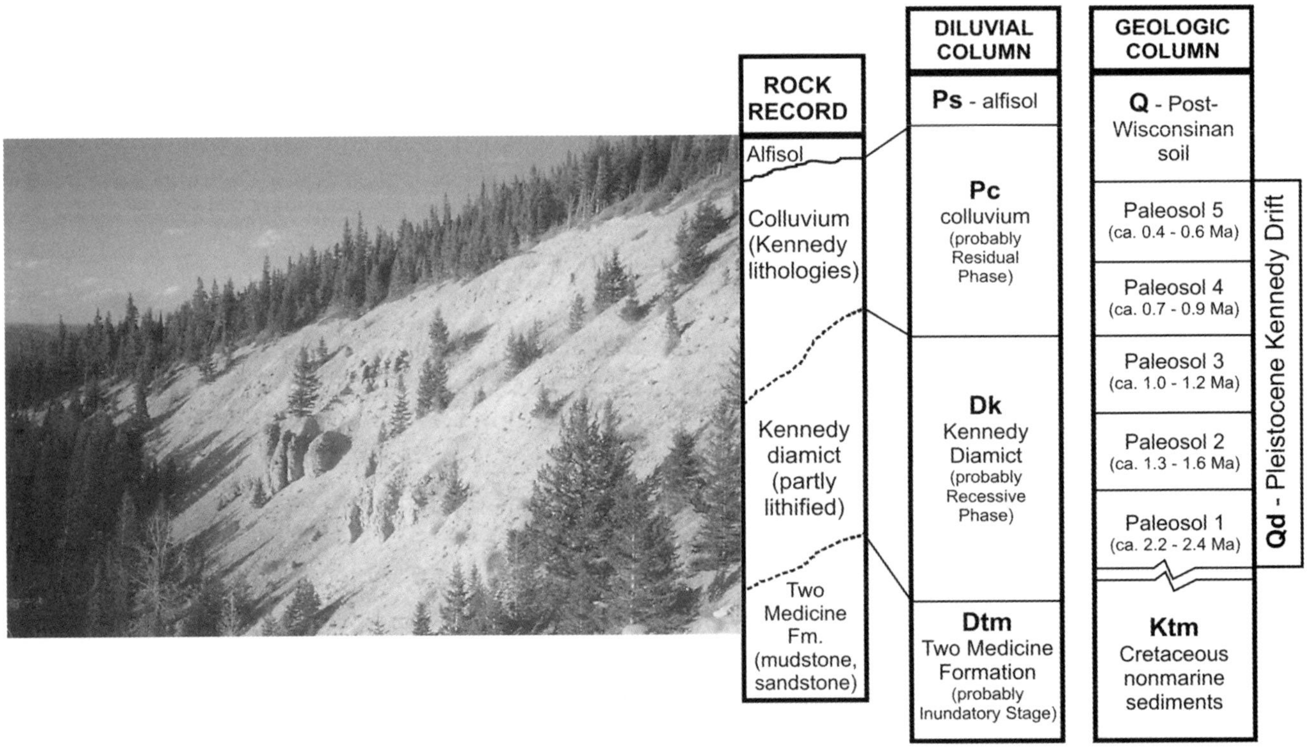

Figure 7. Comparison of uniformitarian and diluvial interpretations of rock record at landslide scarp on the west side of Saint Mary Ridge, Montana. Colluvium mantles Kennedy Formation.

A

System	Series	Group/ Fm.	Description	Unit
Quaternary		Undiff.	Terrestrial clastics in eolian, fluvial, and lacustrine environments.	7
Tertiary	Mio./Plio.	Ogallala		
Cretaceous		Undiff.		6
Triassic	Upper	Dockum	Nonmarine red beds.	5b
Permian	Ochoa		Interfingering beds of salt, carbonate (mostly dolomite) anhydrite, and red beds.	5a
	Guadalupe			
	Leonard			
	Wolfcamp	Undiff.	Deltaic sandstones and shales, shallow marine carbonates & deep basin shale/limestone. Fan deltas of granite wash	4
Penn.	Virgil	Cisco		
	Missouri	Canyon		
	Desmoines	Strawn		
	Atoka	Bend		
	Morrow			
Miss.	Meramec	Undiff.	Shallow marine carbonates; interbeds sandstone, shale	3
	Osage			
Ordovician	Canada	Ellenburger	Shallow marine dolomite	2b
Cambrian			Basal lag sandstone	2a
			Silicic igneous rocks, small mafic portion	1
Precambrian			Igneous crust	

B

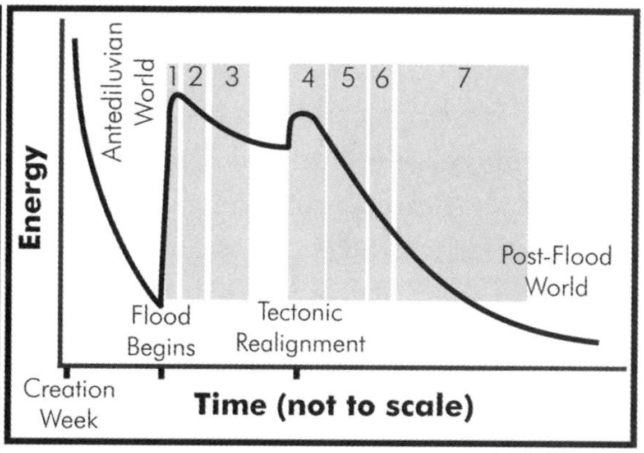

Figure 8. Comparison of interpretations of the Palo Duro Basin, Texas. A=interpretation by column. Moving to the right of A, empirical aspects are emphasized, resulting in a series of genetic units based on bounding surfaces and composition. Dashed lines=unconformities. B=interpretation of these units by changing energy levels associated with the Genesis Flood offers an alternate history. Modified from Reed (2002a).

new clarity that reveals logical incompatibilities of the column with biblical history. A moment's reflection allows one to remember that logical invalidity trumps empirical data or utilitarian practicality. Ironically, diluvial use of the column is itself impractical; significant opportunities to reinterpret the rock record in ways unimaginable to those caught in the straitjacket of Lyellian conformity are lost.

Though the column encapsulates the uniformitarian geologic interpretation of Earth's crust, that interpretation includes the rejection of the Christian worldview in favor of naturalism—a worldview that replaces a reality founded on God's Creation and governance of the universe with an impersonal, uncaring mechanism (i.e., chance), a worldview that substitutes Christianity's confidence in truth granted by God's revelation with an unstable positivism that succeeds only when it pilfers Christian doctrine (Reed, 2001), a worldview that substitutes a meaningful and rich history with a timeframe designed to dismiss the immanent presence of the Creator, and fills its endless ages with pseudo-scientific "just-so" stories. When we consider the vast chasm that lies between naturalism and Christianity, we do not see how creationists can escape the necessity of rebuilding natural history and its derivative stratigraphy from the ground up, as advocated by Whitcomb and Morris (1961).

Creationist proponents of adopting the stratigraphic column believe that they are simply incorporating an empirical construct without any stain of naturalism. We are confident that they decry any influence of that worldview, but we fear that they are missing many connections between the column and the worldview. All creationists should closely examine the assumptions, methods, and conclusions that underlie the column. If they do, we are convinced that they will clearly see the difference between the perceptual rock record and the conceptual column, completely eliminate the all vestiges of evolutionary uniformitarianism from their approach to stratigraphy, and, confidently proceed to develop the diluvial paradigm of natural history within the biblical worldview.

References

CRSQ—*Creation Research Society Quarterly*

CENTJ—*Creation Ex Nihilo Technical Journal* (now *Journal of Creation)*

Ager, D.V. 1993. The nature of the stratigraphical record, third edition. John Wiley and Sons, New York.

Akridge, A.J. 1998. A Flood-based origin of Little River Canyon near Ft. Payne, Alabama. In Walsh, R.E. (editor). Proceedings of the Fourth International Conference on Creationism. Technical Symposium Sessions. pp. 9–20, Creation Science Fellowship, Pittsburgh, PA.

———. 2000. DeSoto Falls in DeSoto State Park, Alabama: Evidence for recent formation? *CRSQ* 36:170–176.

Akridge, A.J. and C.R. Froede, Jr. 2000. Rock spires (pseudo-hoodoos) on the Lookout Mountain Syncline. *CRSQ* 36:216–220.

Austin, S.A. 2004. Nautiloid mass-kill and burial event, Redwall Limestone of Grand Canyon region. *Proceedings of the Fifth International Conference on Creationism.* Creation Science Fellowship, Pittsburgh, PA.

Austin, S.A. and K.P. Wise. 1994. The pre-Flood/Flood boundary: as defined in Grand Canyon, Arizona and Eastern Mojave Desert, California. In Walsh, R.E. (editor). *Proceedings of the Third International Conference on Creationism.* Technical Symposium Sessions. pp. 37–47, Creation Science Fellowship, Pittsburgh, PA.

Barzun, J. 2000. *From dawn to decadence: 500 years of western cultural life.* Harper Collins, New York.

Berggren, W.A. and J.A. Van Couvering (editors). 1984. *Catastrophes and earth history: the new uniformitarianism.* Princeton University Press, Princeton, NJ.

Berry, W.B.N. 1968. *Growth of a prehistoric timescale: based on organic evolution.* Freeman, San Francisco.

Beus, S.S. and M. Morales. 1990. Introducing the Grand Canyon. In Beus, S.S. and M. Morales (editors). *Grand Canyon Geology.* pp. 1–9. Oxford University Press, New York.

Bliss, R.B., G.E. Parker, and D.T. Gish. 1980. *Fossils: key to the present.* CLP Publishers, San Diego, CA.

Brett, C.E. and G.C. Baird (editors). 1997. *Paleontological events: stratigraphic, ecological, and evolutionary implications.* Columbia University Press, New York.

Burdick, C. 1979. *Five minutes with the Bible and science.* Bible-Science Association IX(9):1–3.

Clark, G. 1994. *Historiography secular and religious.* The Trinity Foundation, Jefferson, MD.

Cleland, C.E. 2001. Historical science, experimental science, and the scientific method. *Geology* 29(11):987–990.

Clough, C.A. and L.E. Fredericks. 1978. Creationist science: a challenge from Professor Young. *CRSQ* 15:47–52, 35.

Donovan, S.K. (editor). 1989. *Mass extinctions: processes and evidence.* Columbia University Press, New York.

Erwin, D.H. 1993. *The great Paleozoic crisis: life and death in the Permian.* Columbia University Press, New York.

Fields, W.W. 1976. *Unformed and unfilled: a critique of the Gap Theory.* Burgener Enterprises. Collinsville, IL.

Froede, C.R., Jr. 1995. A proposal for a creationist geological timescale. *CRSQ* 32:90–94.

———. 1997. The Flood event/ice age stratigraphic boundary on the United States Southeastern Coastal Plain. *CRSQ* 34:75–83

———. 1998. *Field studies in catastrophic geology.* Creation Research Society Books, St. Joseph, MO.

———. 2005a. The Tertiary stratigraphy surrounding Americus, Georgia: evidence in support of the young-earth flood framework. *CRSQ* 42:85–90.

———. 2005b. The impact of Hurricane Georges (September 28, 1998) across Dauphin Island, Alabama. *Southeastern Geology* 44(1):45–51.

———. In press. The impact that Hurricane Ivan (September 16, 2004) made across Dauphin Island, Alabama. *Journal of Coastal Research.*

Froede, C.R., Jr. and J.K. Reed. 1999. Assessing creationist stratigraphy with evidence from the Gulf of Mexico. *CRSQ* 36:51–60.

Garner, P. 1996a. Where is the Flood/post-Flood boundary? Implications of dinosaur nests in the Mesozoic. *CENTJ* 10(1):101–106.

———. 1996b. Continental flood basalts indicate a pre-Mesozoic Flood/post-Flood boundary. *CENTJ* 10(1):114–127.

Garton, M. 1996. The pattern of fossil tracks in the geological record. *CENTJ* 10(1):82–100.

Gentet, R.E. 2000. The CCC model and its geologic implications. *CRSQ* 37:10–21.

Gish, D.T. 1985. *Evolution: the challenge of the fossil record.* Creation-Life Publishers, El Cajon, CA.

———. 1995. *Evolution: the fossils still say NO!* Institute for Creation Research, El Cajon, CA.

Gould, S.J. 1997. *Time's arrow time's cycle: myth and metaphor in the discovery of deep geological time.* Harvard University Press, Cambridge, MA.

Hallam, A. 1992. *Phanerozoic sea-level changes.* Columbia University Press, New York.

Harbaugh, J.W. 1974. *Stratigraphy and the geologic timescale, second edition.* Brown Publishers, Dubuque, IA.

Hasker, W. 1983. *Metaphysics: constructing a world view.* InterVarsity Press, Downer's Grove, IL.

Holt, R.D. 1996. Evidence for a late Cainozoic Flood/post-Flood boundary. *CENTJ* 10(1):128–167.

Jablonski, D, D.H. Erwin, and J.H. Lipps (editors). 1996. *Evolutionary Paleobiology.* University of Chicago Press, Chicago.

Jackson, W. 1980. *The mythology of modern geology.* Apologetics Press, Stockton, CA.

Klevberg, P. 1998. The Big Sky Paving gravel deposit, Cascade County, Montana. *CRSQ* 34:225–235.

———. 1999. The philosophy of sequence stratigraphy Part I–philosophic background. *CRSQ* 36(2):72–80.

———. 2000a. The philosophy of sequence stratigraphy Part II—application to stratigraphy. *CRSQ* 37(1):36–46.

———. 2000b. The philosophy of sequence stratigraphy Part III—application to sequence stratigraphy. *CRSQ* 37(2):94–104.

Klevberg, P. and M.J. Oard. 1998. Paleohydrology of the Cypress Hills Formation and Flaxville Gravel. In Walsh, R.E. (Editor). *Proceedings of the Fourth International Conference on Creationism.* Technical Symposium Sessions. pp. 361–378, Creation Science Fellowship, Pittsburgh, PA.

———. 2005. Drifting interpretations of the Kennedy gravel. *CRSQ* 41:289–315.

Lemon, R.R. 1991. *Principles of stratigraphy.* Merrill Publishing Company, Columbus, OH.

Lubenow, M.L. 2004. *Bones of contention: a creationist assessment of human fossils, second edition.* Baker Books, Grand Rapids, MI.

McInnis, D. 1998. And the waters prevailed. *Earth* 7(4):47–54.

Middelmann, U. and A.E. Wilder-Smith. 1980. *Den som tenker må tro* [He who thinks must believe]. Prokla-Media, Ottestad, Norway.

Morris, H.M. 1996. The geologic column and the Flood of Genesis. *CRSQ* 33:49–57.

Mortenson, T. 2004. *The great turning point: the church's catastrophic mistake on geology before Darwin.* Master Books, Green Forest, AR.

Morton, G.R. 1995. *Foundation, fall and flood: a harmonization of Genesis and science*, second edition. DMD Publishing, Dallas, TX.

Myles, D. 1985. *The great waves.* McGraw-Hill, New York.

North American Commission on Stratigraphic Nomenclature. 2005. North American Stratigraphic Code. *American Association of Petroleum Geologists Bulletin* 89(11):1547–1591.

Northrup, B.E. 1977. A critique of the contemporary division/flood model. *Bible-Science Newsletter* 15(12):2–6.

———. 1979. *Five minutes with the Bible and science.* Bible-Science Association IX(2):1–4.

———. 1986. A walk through time: a study in harmonization. In Walsh, R.E. (editor), *Proceedings of the first International Conference on Creationism*, technical symposium sessions. pp. 147–156. Creation Science Fellowship, Pittsburgh, PA.

———. 1990a. Identifying the Noahic Flood in historical geology: Part One. In Walsh, R.E. (editor), *Proceedings of the second International Conference on Creationism*, pp. 173–179. Creation Science Fellowship, Pittsburgh, PA.

__. 1990b. Identifying the Noahic Flood in historical geology: part Two. In Walsh, R.E. (editor), *Proceedings of the second International Conference on Creationism.* pp. 181–188. Creation Science Fellowship, Pittsburgh, PA.

Oard, M.J. and P. Klevberg. 1998. A diluvial interpretation of the Cypress Hills Formation, Flaxville Gravel, and related deposits. In Walsh, R.E. (Editor). *Proceedings of the Fourth International Conference on Creationism.* Technical Symposium Sessions. pp. 421–436, Creation Science Fellowship, Pittsburgh, PA.

Pearcey, N.R. and Johnson, P.E. 2004. *Total truth: liberating Christianity from its cultural captivity.* Crossway Books, Wheaton, IL.

Plantinga, A. 1990. *The twin pillars of Christian scholarship.* Calvin College and Seminary, Grand Rapids, MI.

———. 1997a. Methodological naturalism? *Origins & Design* 18(1):18–27.

———. 1997b. Methodological naturalism: Part 2. *Origins & Design* 18(2):22–34

Price, G.M. 1923. The new geology. Cited in: *Five minutes with the Bible and science*. Bible-Science Association V(8):6 (1975).

Reed, J.K. 2000. *The North American Midcontinent Rift System: an interpretation within the biblical worldview*. Creation Research Society Books, St. Joseph, MO.

———. 2001. *Natural history in the Christian worldview: foundation and framework*. Creation Research Society Books, St. Joseph, MO.

———. 2002a. Reinventing stratigraphy at the Palo Duro Basin. *CRSQ* 39:25–39.

———. 2002b. Time warp I: The Permian-Triassic boundary in the Texas Panhandle. *CRSQ* 39(2):116–119

———. 2003. The geology of the Kansas Basement: Part I. *CRSQ* 40(3):151–163.

———. 2004a. Time warp III: The Proterozoic-Paleozoic hiatus in the northern midcontinent. *CRSQ* 41(1):75–78.

———. 2004b. The geology of the Oklahoma basement. *CRSQ* 41(2):156–167.

———. 2005. The geology of the Timbered Hills Group in Oklahoma. *CRSQ* 42(1):39–66.

Reed, J.K. and C.R. Froede, Jr. 1997. A biblical Christian framework for Earth history research Part III—constraining geologic models. *CRSQ* 33:285–292.

———. 2000. Bible-based Flood geology: two different approaches to resolving Earth history—a reply to Tyler and Garner. *CRSQ* 37:61–66.

———. 2003. The uniformitarian stratigraphic column—shortcut or pitfall for creation geology. *CRSQ* 40:21–29.

Reed, J.K., C.R. Froede, Jr., and C.B. Bennett. 1996. The role of geologic energy in interpreting the stratigraphic record. *CRSQ* 33:97–101.

Reed, J.K., P. Klevberg, C.B. Bennett, C.R. Froede, Jr., A.J. Akridge, and T.L. Lott. 2004. Beyond scientific creationism. *CRSQ* 41(3):216–230.

Robinson, S.J. 1996. Can Flood geology explain the fossil record? *CENTJ* 10(1):32–69.

———. 1997. The Geological Column: a concept foundational to Flood geology. *Origins* 23:14–30.

Ryan, W. and W. Pitman. 1998. *Noah's flood: the new scientific discoveries about the event that changed history*. Simon and Schuster, New York.

Sauer, J.A. 1996. The river runs dry: creation story preserves historical memory. *Biblical Archaeology Review* 22(4):52–57, 64.

Schaeffer, F.A. 1968. *Escape from reason*. In *The complete works of Francis A. Schaeffer—volume I: a Christian view of philosophy and culture*. Crossway Books, Westchester, IL.

———. 1976. *How should we then live?* In *The complete works of Francis A. Schaeffer—volume V: a Christian view of the West*. Crossway Books, Westchester, IL.

Setterfield, B. and H. Setterfield. A snowball earth or…? http://www.setterfield.org/snowballearth.htm (November 20, 2005).

Snelling, A.A., M. Ernst, E. Scheven, J. Scheven, S.A. Austin, K.P. Wise, P. Garner, M. Garton, and D. Tyler. 1996. The geological record. *CENTJ* 10(3):333–334.

Snelling, A.A. 1996. Special symposium: where should we place the Flood/post-Flood boundary in the geological record? *CENTJ* 10(1):29–127.

Titus, H.H., M.S. Smith, and R.T. Nolan. 1979. *Living issues in philosophy*. Wadsworth Publishing Company, Belmont, CA, pp. 217–243.

Tyler, D.J. 1997. Flood models and trends in creationist thinking. *Creation Matters* 2(3):1–3.

Tyler, D.J. and P. Garner. 2000. The uniformitarian column and Flood geology: a reply to Froede and Reed (1999, *CRSQ* 36:51–60). *CRSQ* 37:60–61.

Walker, T. 1994. A Biblical geologic model. In Walsh, R.E. (editor). *Proceedings of the Third International Conference on Creationism*. Technical Symposium Sessions. pp. 581–592, Creation Science Fellowship, Pittsburgh, PA.

———. 1996a. The basement rocks of the Brisbane area, Australia: where do they fit in the creation model? *CENTJ* 10(2):241–257.

———. 1996b. The Great Artesian Basin, Australian. *CENTJ* 10(3):379–390.

———. 2001. Post-Flood volcanism on the Banks Peninsula, New Zealand. *CENTJ* 15(1):96–104.

Whitcomb, J.C. and H.M. Morris. 1961. *The Genesis Flood*. Baker Book House, Grand Rapids, MI.

Wilder-Smith, A.E. 1988. *The creation of life: a cybernetic approach to evolution*. The Word for Today Publishing, Costa Mesa, CA, pp. 189–211.

———. 1993. *Man's origin, man's destiny, second edition*. The Word for Today Publishing, Costa Mesa, CA.

Wilgus, C.K., B.S. Hastings, C.G. St. C. Kendall, H.W. Posamentier, C.A. Ross and J.C. Van Wagoner (editors). 1988. *Sea-level changes: an integrated approach*. Society of Economic Paleontologists and Mineralogists Special Publication 42, Tulsa, OK.

Woodmorappe, J. 1981. The essential nonexistence of the evolutionary-uniformitarian geologic column: a quantitative assessment. *CRSQ* 18:46–71.

———. 1983. A Diluviological treatise on the stratigraphic

separation of fossils. *CRSQ* 20:133–185.

———. 1996. Studies in Flood geology: clarifications related to the "reality" of the geological column. *CENTJ* 10(2):279–290.

———. 1999a. *Studies in Flood Geology, second edition.* Institute for Creation Research, Santee, CA.

———. 1999b. The geologic column: does it exist? *CENTJ* 13(2):77–82.

Authors

John K. Reed, Ph.D., has served in the Creation Research Society as a board member and as the geology editor for the *Creation Research Society Quarterly.* He is the author of five books and numerous articles.

Peter Klevberg, B.S., P.E. is a geological engineer with experience in mineral exploration, geotechnical engineering and hydrogeology. He has published articles about stratigraphy, paleosols, and diluvial deposits in the northern Rocky Mountains.

Carl. R. Froede, Jr., B.S., P.G. is a lifetime member of the Creation Research Society. He is the author of one book and numerous geological articles, many within the Creation/Flood framework.

Forum Discussion

1. In, "First Problem: The Fallacy of the Empirical Modular Column" (paragraph three), you state that the global chronostratigraphic column guides local lithostratigraphic and biostratigraphic columns. Could you give an example of this? (Oard)

Examples are everywhere: logically, the juxtaposition of a global conceptual column with any unexplored section of the crust always results in a "new example" of the column. So, any section outside of 19th century Western Europe illustrates the point since the column was first developed there and was then extrapolated worldwide (see Chapter 2). Turning to more specific examples, how would one identify "missing section" at the Grand Canyon? Without the template of the column, how would we know which formations are "missing" when the physical evidence does not suggest it? Another example is the global imposition of Precambrian chronostratigraphy from the offices of the International Stratigraphic Commission via an arbitrary set of absolute ages, determined by radiometric dating (e.g., base Proterozoic = 2500 Ma). Finally, we take issue with the claim that the column is "global" in the face of unexplored tracts of the crust. Though less egregious than 19th century claims to be global, this mindset confirms its role as a template.

2. In the next paragraph, you say that naturalism has affected lithostratigraphy. Since lithostratigraphy is supposed to be empirical, could you state the basis why you think this is true? (Oard)

The cited paragraph discusses the impact of presuppositions on investigative methods. Even an apparently empirical method like lithostratigraphy is affected. Apart from the philosophical distortions of positivism, lithostratigraphy relies on the assumption of uniformitarianism in at least two ways. First, it presumes slow, gradual deposition evenly spread across time. Otherwise, stratigraphers would not attempt to use the thickness of rock units (assuming uniform deposition rates) to extrapolate age. Furthermore, stratigraphers use modern depositional environments to predict geometry, expected lithologic successions, and expected erosional surfaces. This illustrates the difference between theory and practice; theoretically, lithostratigraphy is just as amenable to diluvial investigation—it just doesn't always work that way in practice. Also, it is worth noting that lithostratigraphy has very little practical influence on the definition of the column when compared to biostratigraphy, geomagnetic stratigraphy, and geochronology.

3. I agree that the column is built on the wrong philosophies of naturalism, evolution, uniformitarianism, and deep time. But what would you say to the claim that, regardless, the right order of strata is found in vertical sequences all over the Earth, except in those areas complicated by overthrusts, reworking, etc.? (Oard)

If you are claiming that the rock record is intrinsically arranged in a succession of global synchronous correlative time periods, we would be forced to disagree. As we stated, it is that assumption that forces interpretation to that conclusion, not data. This is not to say that the rock record does not possess inherent order to a high degree—otherwise stratigraphy would be impossible. One would expect some kind of general global order arising from a global flood, but the real question is whether or not that order is best interpreted by a scheme based on assumptions antithetical to biblical history. Given the time frame of the Flood, and the resulting short-term variations in hydraulic, tectonic, and thermal energy over time and space, as well as a rapidly changing base level in both time and space, one would expect that parameters other than time would be best suited to unlock the secrets of the rock record. Though the step from inherent order to the geologic column may seem short when viewed from a distance, a closer look reveals it to be a rather large chasm.

4. You have argued that a synthesis of local columns "implies an overarching plan guiding the assembly of the various parts." Creationists who accept the Column do not argue for an overarching plan, but for a methodology involving principles of correlation. How do you respond to this "bottom-up" approach? (Tyler)

We support a "bottom-up" approach, but consider the methodology of using the column as a global template to be "top-down." Of course, the process of defining any methodology is inescapably non-empirical (for both uniformitarians and diluvialists), as we have taken some pains to explain. Fundamental to your proposed column is the assumption of an idealized vertical succession of time periods that are the same everywhere, and are everywhere correlative. That "top-down" approach is fundamental to a global column—whether made by consciously by uniformitarians or unconsciously by creationists. We question whether that assumption can be integrated with biblical history. We strongly encourage a "bottom-up" approach (see Chapter 3, "Towards an Empirical Stratigraphy), but the column does not offer it. Part of the problem is scale; as we stated, we encourage diluvial investigations to the limits of correlation, but large-scale correlations always seem to rest on biostratigraphy with its assumed evolutionary progression, geomagnetic stratigraphy (calibrated by radiometric dating), and geochemical measurements tied to Milankovitch cycles, etc.

5. You refer to "contrary empirical evidence," but most of the argument presented involves "conceptual aspects, especially its presuppositions." Please indicate the empirical evidence that contradicts the synthesis of local columns. (Tyler)

Before doing so, it is worthwhile to point out that presuppositional and logical errors are often less evident, but are always more deadly than empirical mistakes. Therefore, we are relieved that our emphasis on those dangers was clearly communicated. Defining empirical evidence that contradicts "the synthesis of local columns" depends on both the scale and style of the synthesis. Given the context of this discussion, we assume you mean a global synthesis into synchronous correlative time periods. In that case, in addition to the numerous examples cited in Chapters 6 and 7, we see the modern multiplication of "markers" as an empirical weakness, much as the continual stream of "new and improved" radiometric methods (usually to correct defects in of older ones) highlights the weaknesses of that pursuit. Obviously, if widespread correlation was straightforward, then there would be no need to multiply new correlation criteria. Anytime multiple markers do not agree, another piece of evidence emerges. Any of the numerous "corrections" made to the column over the years (e.g., reassigning a formation based on changing fossil ranges) would represent contrary empirical evidence. The column's ability to absorb all of these changes and still maintain its basic structure supports our contention that, at heart, it is not an empirical construct, but a trans-empirical template.

6. You write: "In summary, the severely limited duration and the catastrophic nature of the biblical Flood seem to preclude the possibility of an ideal global vertical succession of laterally isochronous strata." How do you handle the charge that your preconceptions of 'what the Flood was like' has become a dogma that stops you recognizing patterns in the rock record? (Tyler)

First, we would note that our preconceptions are probably no more or less dogmatic than anyone else's. Next, recognizing that possibility, we make it a practice to be "up front" with our assumptions and to avoid investing our forensic speculations with the unwarranted authority of either the Bible or scientific demonstration. To the extent that our preconceptions coincide with Scripture, we gladly embrace them as well as the boundaries it provides in addressing the rock record. Second, we do recognize a variety of inherent patterns in the rock record—they are the basis for our advocacy of an empirical stratigraphy. However, we do not accept the artificial pattern of global synchronous correlative time periods as the sole authoritative template for interpreting those patterns. There are also patterns that indicate catastrophic, regional erosion, deposition, and tectonism. Some see our approach only negatively (i.e., get rid of stratigraphy), but in reality, it is both innovative and liberating for diluvial geology, offering the means by which diluvialists can retake the rock record from its long captivity to uniformitarianism.

7. You have the commendable vision of developing "the diluvial paradigm of natural history within the biblical worldview." This means that we must do more than describe rocks and show that they could be emplaced quickly. It means we must do more than offer interpretations of how they fit into biblical history. It means we must seek to reconstruct geological histories that make sense of the rock record. How can this be done without developing principles of correlating local columns? (Tyler)

We are pleased that you recognize the scope of the project facing diluvial geology. Investigation of the rock record and the centrality of biblical history in constraining interpretation are both important foundations of diluvial geology; however, we agree that they are only the beginning. Progress will most certainly include defining principles of correlation. But sometimes when something new is being defined, it helps to state what it is not, as well as what it is. Therefore, in the interest of clarity, we overtly reject any such principles that rest on assumptions of deep time, evolutionary progression, and uniformitarianism. Given the expected hydraulics of a one-year global flood, we also remain skeptical of the presence of global stratigraphic markers ubiquitous in all four dimensions, but we stand ready to be convinced. Also we lean towards environment being more significant than time in stratigraphic interpretation. This suggests that tectonic, sedimentological, chemical, and hydraulic understanding may be of greater value than simply pigeonholing strata by time. In answer to your final question, the second footnote in Chapter 8 lists published projects providing examples of our approach to diluvial geology.

nine

Diluvial Perspectives on the Geologic Column

John K. Reed
Michael J. Oard

Regardless of how diluvial geologists decide to apply the geologic column, they stand together against the uniformitarian stratigraphy of Lyell and his disciples. Increasingly desperate attempts by uniformitarians to rest one foot firmly on uniformitarianism while standing with the other on neocatastrophism demonstrate the theoretical failure of this paradigm. Its practical fall cannot be far behind. The scientific revolution to replace it will probably be both conceptual and technological. Conceptually, it will see the replacement of an outdated uniformitarianism with catastrophism. From a diluvialist perspective this trend will be reflected in biblical catastrophism. Technologically, the overthrow of uniformitarianism will be reflected in the increasing empirical comprehension of the crust and its constituent units as they become better understood in both geometry and composition.

Diluvial geologists are positioned to play a significant role in this development, especially within the theoretical realm. Only biblical catastrophism offers a coherent and consistent method of investigating geologic history. Reinstating biblical catastrophism as the default paradigm of natural history will first require defining core principles for natural history within a biblical framework. These must move beyond an outmoded positivism and integrate biblical authority with forensic innovation without sacrificing either. An immediate benefit of that effort will be the restoration of appropriate levels of confidence for historical models: no longer will science be forced to perform tasks better suited to history or philosophy—tasks at which it has proven so inept. A biblical reaffirmation of the unity of truth should also bring about a closer marriage between theoretical and practical geology.

These changes will probably not occur in the same manner as the 19th century uniformitarian revolution because the public no longer shares its naïve Enlightenment optimism. Furthermore, the empirical basis for stratigraphy is much greater; a few diverting generalities from the Alps, Niagara Falls, or the English countryside cannot as easily hide interpretive aspects of historical conclusions. As Christians, we should welcome the opening of long-closed windows that have hidden the bias of those who use the methodological objectivity of science to make the absurd claim of metaphysical neutrality. A willingness to openly admit and discuss philosophical foundations is an advantage for diluvialists. Proponents of uniformitarianism are quick to criticize the imposition of "religion" into science, but remarkably reticent to admit the not-inconsiderable role of their own beliefs and assumptions. Both sides should be ready to discuss their worldview in all its aspects—theological, philosophical, historical, and scientific.

Most of that discussion is ongoing as this book has examined the relatively narrow debate within diluvial circles about the nature and role of the geologic column. It has shown many areas of agreement between the different perspectives and has highlighted those particular issues that have yet to be resolved. Given the current state of affairs, it would be well worth summarizing those two areas.

Before doing so, however, there is a clarification worth examining—that between common ground and middle ground. One of the baleful effects of our modern civilization has been the attempt to muddle truth and opinion. Within that context, debate often becomes subjective or is claimed to be no more than subjective, and victory is claimed by those "more reasonable" who are able to find "middle ground." In fact, some see that as the only eminently reasonable and sane path to pursue. To the extent that differences are purely subjective, this mediatory approach is to be applauded. However, in objective debate, the pursuit of truth demands the exclusion of those things that are not. Thus, to the extent that our differences are subjective matters of taste, then middle ground is desired: to the extent that propositions of particular positions are either true or false, and contrary, middle ground is neither possible nor desirable. Instead, we need to affirm *common* ground and strive to enlarge that area through evidence and valid argument.

Common Ground

Necessity of Biblical History

The best understanding of the past comes from written records of events. The Bible presents a written record of ancient history, but it forces a natural divide because the implications of its accuracy extend far beyond the realm of history. No other document offers such a clear, specific, validated eyewitness account of the beginning of the universe or of man's earliest days. In addition to its reliability from the point of view of historiography, the Bible makes the claim in many places to be the account of God—the one Being who possesses the depth and integrity of knowledge to provide absolute historical reliability.

Modern thought since the Enlightenment has been united in its rejection of both God and the Bible as God's revelation of history. Geology has played a crucial role in the development of that aberration. Since the 19th century, it has served as a wedge of doubt in the Christian understanding of the past. By positing "prehistory" as both a time before any written record and as the bulk of Earth's past, it has simultaneously cast the Genesis accounts as wildly postdating the world's origin and early history—thus irrelevant—and has replaced it with the account "written" in the rock record. Thus, the traditional understanding of history as a discipline has been displaced by the mistaken concept that the historical information accessible by scientific means is relatively unlimited.

The recent rise of creationism and diluvial geology seeks to reverse that trend and all the participants of this symposium stand together in affirming the Bible as God's revelation to man, accurate in its historical record, and therefore a necessary foundation to any understanding of origins or ancient history. In addition to serving as a foundation, the Bible supplies boundaries; it functions as a primary truth test for any model, paradigm, or theory about natural history.

Furthermore, we affirm that the Bible provides the intellectual foundations for any natural history and natural science, based on its presentation of the nature of the cosmos, the nature of man, and the relationships among God, man, and the rest of the created order. In addition, our motivation for pursuing truth about Creation and the Flood through the medium of the earth sciences—uncovering and proclaiming the glory of God—is itself derived from the Bible. Therefore, we all assert our dependence on the Bible for its historical accounts, its intellectual foundation of other methods of investigation, and our motivations for pursuing truth.

Commitment to Empirical Stratigraphy

In addition to fidelity to Scripture, each of the contributors of this book affirms a strong commitment to the empirical investigation of the rock record though the appropriate methods of stratigraphy and other related branches of the earth sciences. All of us are geologists by avocation and most of us are by education and vocation.

While the Bible is authoritative and constrains speculation, it sometimes lacks detail from the perspective of geology. We are convinced that a strong effort to comprehend the rock record through field investigation, the study of relevant reports in the professional literature, and experimental studies will result in a wealth of information about the bare outline of events so tantalizingly presented in thc short accounts of the early chapters of Genesis.

An empirical stratigraphy presumes that there are discernable characteristics of the rock record that can be proper objects of investigation. These would include concepts such as internal order within beds and within successions of beds, regularities in lateral and vertical successions of lithology and fossil content, and other similar properties that allow unique classification and which might allow correlation to some reasonable scale. These stratigraphic properties can be examined by using a multitude of methods: field observation of outcrops, subsurface geophysical measurements, and laboratory investigation. Though we may disagree on some details of the parameters of such investigation, we are all committed to that basic task.

Catastrophism

One of the foundations of diluvial stratigraphy rests upon the recognition that the 19th century proposal of uniformitarianism was a dead end for geological research. The rock record and the Bible both bear testimony to upheavals in the past, far different from the geologic processes that we observe today in scale, intensity, and rate. Furthermore, we recognize that the quantitative difference between the events of the Flood and those observed in the present may easily result in qualitatively distinct processes that have no obvious parallel in today's world.

Therefore, we join an increasing number of our secular colleagues who recognize that the rocks clearly speak to catastrophic erosion, tectonism, igneous activity, sedimentation, and diagenetic and metamorphic alteration. The straitjacket imposed on geology by Hutton and Lyell must be removed to make way for the conceptual revolution that is stirring within modern stratigraphy. We are also united in our opposition to the attempts of our secular colleagues to integrate uniformitarianism and catastrophism—sometimes under the guise of "actualism" as a practical alternative. However, the theoretical foundations cannot be ignored so easily. Geologists in the 19th century at least understood that the two approaches to natural history could be united about as readily as fire and water! Furthermore, Earth's history is not marked by a succession of small-scale disasters; it is dominated by the singular events of Creation and the Genesis Flood. We join in affirming that the recognition of these events must be foundational to any investigation of history.

Rejection of Evolution

Modern stratigraphy dates, orders, and correlates strata primarily on the basis of an assumed evolutionary order reflected by fossil content. Our emphasis on a more empirical stratigraphy and our understanding of biblical history combine to lead us to the rejection of the possibility of Neo-Darwinian or Punctuationist forms of macroevolution. Biotic remains in the rock record reflect life created in the beginning by God that has exhibited the marvellous ability to adapt within their created kinds to the changing ecological pressures manifested throughout time. No new "kind" has arisen by mutation and natural selection, and life reflects the long-term thermodynamic degradation described theologically by the Edenic curse. The fossil record demonstrates death and decay; not ongoing evolutionary advances in order and complexity.

Reduced Timescale

If the rock record formed catastrophically, the concept of "deep time" is irrelevant to its interpretation. Biblical history combines with the logical necessity of catastrophism to lead to an affirmation of a young age for Earth and its inhabitants. The first victory of Enlightenment geology was a widespread scepticism of the historical accuracy and detailed chronological information found in Scripture. Those efforts persist today; ridicule of Archbishop James Ussher serves as a type of litmus test for intellectual respectability—even when a cursory examination of Ussher's work reveals a mind far beyond the reach of his critics. While Christians may honestly question interpretive details of biblical history, there is no doubt that it unambiguously teaches that Earth is thousands, not billions of years old.

This has profound effects on diluvial geology. First, the methods that purport to show a history over billions of years are rejected. Foremost among these are the isotopic dating techniques. Second, the significance of geomagnetic and geochemical markers in the rock record moves away from long ages and presents opportunities for innovative research, since they can no longer be ascribed to astronomical perturbations or slow and uniform plate tectonic motion. Furthermore, we affirm that the bulk of the geologic column—the Precambrian—has not been correctly defined in a stratigraphic sense, as stated in the most recent comprehensive publication of the International Commission on Stratigraphy.

> Due to the fact that most Proterozoic and Archean rocks lack adequate fossils for correlation, a different type of boundary definition was applied for subdividing these eons into eras and periods.... For these two eons, the assigned boundary, called a Global Standard Stratigraphic Age (GSSA), is a chronometric boundary and is not represented by a GSSP [Global boundary Stratotype Section and Point] in rocks, nor can it ever be (Gradstein et al., 2004, p. 26, brackets added).

Any subdivision of the Precambrian must rely on as yet undiscovered or explicated criteria to make it a meaningful historical record. Tyler (Chapter 5) has proposed that a crustal catastrophe at the top of the "Hadean" might provide one such marker. However, the stratigraphy of the Precambrian remains relatively undefined when compared to overlying rocks.

Areas to be Resolved

Although there is broad agreement between the authors contributing to this book, there are also many unresolved issues (Table I). This is not surprising: those advocating diluvial geology are a pronounced

Issue	Tyler	Silvestru	Oard	Reed et al.
Science and natural history	natural history is a subset of science		natural history is a subset of of history; science is its primary forensic tool	
Terminology	geologic column is synonym for rock record	no explicit distinction between geologic timescale and rocks	column is generally synonym of rock record	need two terms: *rock record* = empirical; *column* = interpretive
Rock units vs. time units	rocks can be grouped into global time units	rocks can be grouped into global time units	rocks can generally be grouped into global time units	time is not adequate global parameter for rock units
Global correlation	time units same all over earth	time units same all over the earth	largest time units are generally the same globally	global correlation unlikely
Order in the rock record	global order seen in fossils and other markers	general global order seen in a variety of markers	general global order; local exceptions; fossils are a general guide to order	order less likely with increasing scale; most markers local or regional
Deluge and rock record	Flood = Archean to Mississippian	not specified	Flood = most of Phanerozoic; some Precambrian	Flood = most of rock record; most seds. & some basement
Status of Cenozoic	long after Flood	not specified	can be early, late, or post-Flood	time units are not meaningful
Duration of Deluge	"mabbul" = 40 days	not specified	"mabbul" refers to entire 371-day Deluge, not its first 40 days	
Distribution of catastrophism	first day = severe; ongoing for 40 days; less so afterward	not specified	weighted to first 150 days; powerful late erosion	rapid rise and slow fall; tectonism occurs over duration
NT references to Deluge	NT references demand destruction on first day	New Testament references to Deluge do not require biologic or geologic destruction in one day		
Geological scope of Flood	obliterated crust and antediluvian fossils	not specified	crust only fractured, deformed & eroded; many organisms preserved as fossils	
Origin of fossils	fossils from Flood survivors, progeny	most fossilization occurred during the Flood; most fossils are the remains of antediluvian organisms		
Postdiluvial markers	first appearance of terrestrial animals	not specified	postdiluvial markers vary; mainly found in the scope and scale of geologic processes	
Postdiluvial catastrophes	column shows many large postdiluvial catastrophes	catastrophes occurring after the Flood are dramatically limited in both scope and scale when compared to Flood		
Survival of marine life	most killed, remnant in ocean refugia; they repopulate seas	not specified	marine extinction and fossilization primarily during the Flood	

Table I. Topics for further debate.

minority, only beginning to explore the potential offered by this exciting paradigm.

Niche of Natural History and Role of Science

Like many foundations, this issue (Table I) remains largely hidden and may appear unrelated to disputes over the geologic column. Chapter 3 provides some background demonstrating the connections among the disciplines of science, history, and natural history, as well as their role in stratigraphy. The key point of disagreement among the authors of this book is over the extent to which science sets the ground rules for that debate. Most of the authors believe that since the column is a topic of stratigraphy, itself a discipline of geology, and because geology is a scientific enterprise, then resolution (pro or con) of the relationship of the column to diluvial geology is a matter for scientific investigation. That is also the majority opinion of modern science.

By contrast, the minority believes that the study of unique past events is history—by definition—and that natural history is a subset of history rather than science. They subordinate the role of science to history by further asserting that natural history is a "mixed question"—one in which multiple disciplines cooperate to answer intellectual questions that cannot be completely resolved by one area of knowledge alone. They note the necessary relationship of the Bible to diluvial geology, and understand that relationship alone demands a mixed question approach. They also note that the positivism that permeates western culture, and occasionally trickles down into even creationist thought, is an epistemological error that cannot coexist with a biblical worldview.

As with all foundational issues, the relevance of a philosophical understanding of natural history may not be immediately apparent, but if diluvialists do not understand the "rules of the game," they risk conceding the "home field advantage" to those advocating a naturalist view of history that automatically precludes Creation or the Flood. Therefore, proper definitions of science, history, and natural history, and the relationships among these disciplines, the Bible, and its derivative worldview must be made clear to support whatever conclusions diluvialists draw from their study of stratigraphy and the geologic column.

Terminology

In Chapter 8, Reed, Klevberg, and Froede suggest a clarification of terminology to distinguish between the perceptual "rock record" and the conceptual "geologic column," (or "geological timescale" per Silvestru). Tyler uses the two terms synonymously; Silvestru makes no explicit distinction between the geologic timescale and the rock record; and Oard believes that the column is generally synonymous with the rock record. The importance of this distinction in terminology is twofold: (1) are there non-empirical, interpretive facets of the column, and (2) if the column does possess both empirical and interpretive facets, do the conceptual facets drive interpretation or do the empirical aspects of the rocks lead inevitably to stratigraphic conclusions?

Whether or not people see any importance to this issue depends largely on their view of the previous questions about natural history. If stratigraphy is seen to include extra-scientific axioms and methods, then such a distinction is helpful. If stratigraphy is "pure" science, then perhaps it is not helpful. However, it is hard to escape the conclusion that stratigraphy has been strongly influenced by philosophical, theological, and historical conclusions, and it might then be wise to discriminate between empirical and non-empirical facets.

Nature of Rock Units and Time Units

Are rock units (formations, groups, etc.) chronometers? Do they represent periods of time that can be distinguished from one another on a global basis? Uniformitarians believe that they do because that is the foundation for modern stratigraphy; indeed, some conclude that rock and time are virtually identical:

> We consider that the practice of Chronostratigraphy today defines the time framework of Geochronology, because intervals of geological time are now being precisely defined within rock successions by GSSPs [Global boundary Stratotype Section and Point]. The effect of this is that Chronostratigraphy and

> Geochronology should become one and the same discipline... (Gradstein et al., 2004, p. 41, brackets added).

Tyler, Coffin, Silvestru, and Oard all affirm that rock units can be classified and correlated globally as time units. None of these authors affirm the deep time of their uniformitarian colleagues, and Oard is suspicious of the level of resolution that can be obtained, restricting it generally to the level of Era (i.e., Paleozoic, Mesozoic, and Cenozoic). Therefore, the diluvial alternative that embraces the column demands that its timescale be compressed by many orders of magnitude.

In contrast, Reed, Klevberg, and Froede assert that the Flood's (1) rapidity of sedimentation, (2) catastrophic intensity, and (3) local and regional variations in erosion, sedimentation, and tectonism preclude the possibility of rock units representing an orderly arrangement of globally-synchronous time periods. Therefore, they recommend that stratigraphers consider other parameters around which to classify and correlate rock units. Because global isochronous time periods are the basis for the column, this issue is at the heart of this debate.

Possibility of Global Order and Methods of Correlation

Another aspect of the relationship between rock units and time units is whether or not potential time-rock units can be correlated everywhere throughout the crust. Uniformitarians are confident that the top of the Cretaceous section in England was deposited at exactly the same time as the top of the Cretaceous section in North Africa, North America, Northern India, and the North Slope. Tyler, Coffin, and Silvestru agree that rock can be correlated by aligning the time of their formation, which will be the same globally. They expect to find global order in fossil contents (Tyler) or other markers (Silvestru). Oard restricts the possibility of global correlation to the level of Era and does not consider it to be absolute in all cases.

In contrast, Reed, Klevberg, and Froede do not see a diluvial basis for systematically connecting rock units on the basis of their time of formation both vertically and laterally through the rock record. Neither do they accept stratigraphic markers for doing so as being: (1) global (i.e., magnetic or chemical markers) or (2) congenial to the diluvial paradigm (i.e., successions of faunal assemblages). They also question the conclusion that particular phenomena are "global" in the sense that they actually occur everywhere in the crust, when in reality they are simply found in widely scattered areas and extrapolated as ubiquitous. They consider order and correlation less probable as scale increases from local to global. Criteria for the global nature of markers (if any) should be developed more carefully.

The Flood and the Rock Record

If the vast majority of diluvialists agree that a significant percentage of the rock record can be attributed to the Genesis Flood, the question naturally arises as to which part of the record is to be attributed to the Flood. What defines the base of the Flood deposits or eruptives and what defines the top? Can the same criteria be used everywhere? What about intrusive bodies? What about structural features? The authors of this book provide a range of answers—illustrating another area needing further study by creationists.

Although Silvestru does not specify a range, he clearly implies that the vast majority of the rocks resulted from the Flood. Oard classifies most of the Phanerozoic and most, if not all, Precambrian sedimentary rocks as Flood deposits. Reed, Klevberg, and Froede believe that almost all of the Phanerozoic rock record resulted from the Flood, as well as some Precambrian rocks—both igneous and sedimentary. However, they differ from the other authors by advocating a local to regional determination of such boundaries empirically, rather than by the global assignment of particular time periods of the column. Oard also supports this approach. Tyler advocates a different point of view. He assigns all

Precambrian crustal rocks to the Flood, as well as Phanerozoic rocks up to approximately the Mississippian. Rocks chronostratigraphically classified as younger than Mississippian would be the result of post-Flood processes.

Therefore, there are two major differences among the authors: (1) whether or not all Precambrian crystalline rocks are a result of the Flood, and (2) whether or not any Phanerozoic rocks above the Mississippian are a result of the Flood. Resolving these questions would better define the extent to which the rock record and the Flood overlap. Derivative issues include the disposition of Precambrian sedimentary rocks and Phanerozoic plutonic bodies.

Status of the Cenozoic

Related to the question of what part of the rock record reflects Flood processes is the specific issue of the Cenozoic. Tyler recognizes the Cenozoic as a series of post-Flood deposits with associated igneous, metamorphic, and tectonic features emplaced later in history. Silvestru does not specify the disposition of the Cenozoic. Oard recognizes the Cenozoic as a discernable and correlative time-rock unit, but varying significantly in time from early- to post-Flood, depending on the location. Reed, Klevberg, and Froede do not accept "Cenozoic" as a global isochronous correlative time period or time-rock unit. However, they recognize that varying quantities of rocks currently defined as Cenozoic resulted from the Flood, depending on the location.

Distinct "Hypercatastrophism" within the Flood

Though the biblical chronology of the Flood clearly sets the length of time from which Noah entered the Ark until Noah exited the Ark—371 days, Tyler proposes a distinct period—the first forty days—that was dramatically more destructive and catastrophic than the remainder of the Flood. Oard explicitly recognizes that energy expended during the Flood was probably weighted to its onset, but does not agree with Tyler's concept of total crustal destruction on the first day or a specifically-elevated level of geologic activity within the first forty days. Silvestru does not specify energy variations during the Flood, nor do Reed, Klevberg, and Froede, although Reed and Froede (Reed et al., 1996) have proposed a graph showing the distribution of geologic energy over the 371 days of the Flood. In contrast to Tyler and Oard, that graph shows a second energy peak at approximately the middle of the Flood.

A partial resolution of this question can be supplied by Hebrew language scholars. Tyler applies the term, "mabbul" to only the first forty days, and in doing so, infers a much greater degree of destruction during that time period. If Hebrew language scholars can determine whether "mabbul" applies to the first forty days or to the entire 371-day event, then the biblical basis for Tyler's position will be either strengthened or undercut. The *Creation Research Society Quarterly* will welcome and publish such an analysis.

Interpretation of Related New Testament Passages

Tyler suggests that many creationists have missed the relevance of New Testament references to the Flood. Specifically, he notes that comparative analogies of the Flood to both the Second Coming and to the destruction of Sodom and Gomorrah demand an instantaneous destruction of the crust on the first day of the Flood. Furthermore, he notes that Matthew 24:38–41 and Luke 17:26–35 strongly support the total destruction of the crust on the first day.

Although the other authors do not specifically address these verses in their chapters, none of them agree that the passages demand Tyler's chronology of the Flood, defaulting to the explicit chronology supplied in Genesis 6–8. As with the previous issue, this is one that Bible scholars can resolve, and so we request that specialists in these areas assess Tyler's textural arguments and determine the validity of his interpretation. The *Creation Research Society Quar-*

terly will welcome and publish such an analysis.

Geological Scope of Flood

For the past two hundred years, diluvialists have struggled against various versions of a tranquil flood—first proposed to relieve Christians of the responsibility for addressing the 19th century arguments for uniformitarian geology. Therefore, it seems strange that diluvialists would be faced with the question of whether or not a Flood model is "too catastrophic." However, that is a disagreement brought out by this volume. Tyler proposes that the scope of the Flood included the complete destruction and renewal of Earth's crust. Although there are similarities between this proposal and Catastrophic Plate Tectonics (CPT) which advocates the subduction and renewal of oceanic crust, Tyler's concept is even more comprehensive, and his timing differs from the CPT model.

Oard, Reed, Klevberg, and Froede reject a complete crustal overturn, believing instead that although the crust was deformed, intruded, and severely eroded during the Flood, that it was never destroyed. Thus, many crystalline rocks are the remnants of the Creation week and (possibly) antediluvian processes. One criticism of extremely catastrophic models such as Tyler's proposal and CPT is the problem of excess heat in the hydrosphere and atmosphere resulting from the exposure of mantle temperatures at those interfaces. Thus, studies quantifying heat flow during the early Flood by physicists and engineers may be able to help constrain the level of catastrophism consistent with the Bible, the rock record, and the survival of life in the oceans and aboard the Ark.

Origin of Fossils

Related to the catastrophic intensity of the Flood is the question of the origin of fossils found in the rock record. This is an old question: the issue that originally stimulated Nicholaus Steno (Chapter 2) to study geology in the 17th century. As a consequence of Tyler's proposal for complete crustal destruction, the only antediluvian life that would have survived the onset of the Flood would have been those in the Ark and organisms located in selected deep-water marine "refugia." Therefore, the fossils found in the rock record (most in the Phanerozoic) would have come from survivors of the Flood reproducing and migrating to replenish the planet. Also, given Tyler's end point for the Deluge, fossils found in rocks higher than Mississippian would almost undoubtedly represent post-Flood (i.e., off the Ark) generations of fauna and flora.

Although Silvestru does not specify his view, all of the other authors disagree with Tyler's proposal, attributing the bulk of the fossil record to those creatures alive at the time of the Flood, especially all of the air-breathing creatures not taken into the Ark. Therefore, for example, while Tyler views Mesozoic dinosaur fossils as post-Flood both as organisms and as fossils, the others would affirm that the organisms were predominantly antediluvian, and the fossils resulted from the work of the Flood itself.

Post-Flood Markers

A further consequence of the differing correlations between the Flood and rock record is found in determining what features of the rock record may be reliably used to distinguish between Flood and post-Food sediments. Tyler posits the complete destruction of all air-breathing life outside of the Ark on the first day of the Flood and sets the latest exit from the Ark as Mississippian. Thus in his model, the first appearance of terrestrial air-breathing animals in the fossil record marks post-Flood sedimentation.

While the other authors do not specify such markers, Oard, Reed, Klevberg, and Froede would point to local examples of a change in the scale and scope of erosion, sedimentation, and tectonism, to mark the end of the Flood. Furthermore, Oard has published frequently on a post-Flood Ice Age; thus, any Ice-Age deposits would be seen by him as clearly post-Flood.

Post-Flood Catastrophes

Yet another consequence of the various views of when the Flood ended relative to the rock record is the evidence throughout the rock record of catastrophism, and of rapid, regional erosion, deposition, and tectonism well up into the Cenozoic. Most of the authors in this book take this evidence as pointing to the ongoing work of the Flood. However, Tyler's view of a mid-to-late Paleozoic ending for the Flood demands that catastrophic activities later in the rock record were post-Flood in nature. Thus, studies arguing for or against post-Flood catastrophes of sufficient scale and intensity to produce the rocks attributed to the later Paleozoic, and all of the Mesozoic and Cenozoic eras will be of significance to this argument. The limited timeframe between Noah and the emplacement of widespread archaeological artifacts, apparently restricted to the uppermost strata, the relationship of the Ice Age to post-Flood catastrophism, and the constraints of the divine promise in Genesis 9 are all relevant topics that should be thoroughly investigated.

Survival of Marine Life

A final topic for further study may be found in the different views of the continuity of marine life during the Flood. While all the authors agree that large numbers of marine faunas were destroyed by the Flood, Tyler asserts that only those representatives surviving in open-water "refugia" survived annihilation, forming the basis for recolonization of all marine waters, both during and after the Flood. The other authors believe that although many antediluvian marine organisms were killed during the Flood, many were preserved as fossils, largely in those portions of the rock record classified as Phanerozoic. Some marine life survived the catastrophe and became the progenitors of life in our modern oceans. Thus, the 3-dimensional distribution of marine organisms in the rock record, irrespective of purported age, should be of interest in determining if the distribution of these creatures suggests one view over the other.

Conclusion

If nothing else, differing diluvial perspectives on the geologic column illustrate that the recovery of a biblical natural history and a related diluvial stratigraphy are early in their development. Compared to the rapid breakthroughs of the mid 19th century, this may seem slow. However, there are several mitigating factors: an overwhelming uniformitarian predominance in manpower and resources, and a vastly greater empirical database requiring interpretation. Thus, a certain amount of disagreement is to be expected.

In spite of the differing views regarding the nature and work of the Flood, diluvialists firmly agree on several foundational principles. Uniformitarian geology and Darwinian biology are tied to an unworkable worldview of secular naturalism—a worldview which exhibits logical inconsistencies both between its own axioms and conclusions and also with a biblical worldview. Therefore, that worldview of secular naturalism must be set aside. As a consequence, so must the concepts of evolution, uniformitarianism, and deep time. In stratigraphy, these principles will have profound effects. Faunal progression which has historically supplied the "clock" of chronostratigraphy can no longer be considered reliable. Nor can the uniformitarian demand for modern depositional analogues at uniform rates through time. With a return to catastrophism, the rock record must either be seen as the record of a much shorter time period (biblical diluvialism) or a record so sparse that its evidentiary value is nearly meaningless (secular neocatastrophism). Since the uniformitarian view is firmly rooted in an epistemology of pure empiricism, the latter possibility is just as damaging to its survival as the former.

Turning back inward, diluvialists should welcome the development of competing models of natural history. As long as each one is subjected to rigorous analysis by biblical textural studies, logical tests, and empirical constraints, the competition should prove healthy in the long term. Our generation may not see unification around a single model, but we affirm the piecemeal progress that science typically affords. So while we are left with all issues yet unresolved, we are at least left with a clearly defined set of problems that encapsulate the differences presently existing within Flood geology. Hopefully this process of clarification will itself lead to a better understanding of the tasks facing us as well as to the advancement of a framework for geology that will prove to be theologically satisfying, scientifically stimulating, and practical in its theoretical conclusions.

References

Gradstein, F.M., J.G. Ogg, and A.G. Smith. 2004. Chronostratigraphy: linking time and rock. In Gradstein, F.M., J.G. Ogg, and A.G. Smith (editors) *A geologic timescale 2004*. Cambridge University Press, Cambridge, UK.

Reed, J.K., C.R. Froede, Jr., and C.B. Bennett 1996. A biblical Christian framework for earth history research: part IV the role of geologic energy in interpreting the stratigraphic record. *CRSQ* 33(2):97–101.